AF540866

DPH MATHEMATICS SERIES

TEXT BOOK OF SAMPLING AND ATTRIBUTES

By

A.K. Sharma

DISCOVERY PUBLISHING HOUSE

NEW DELHI-110002

First Published-2005

ISBN 81-7141-954-2

Published by

DISCOVERY PUBLISHING HOUSE
4831/24, Ansari Road, Prahlad Street,
Darya Ganj, New Delhi-110002 (India)
Phone: 23279245 • Fax: 91-11-23253475
E-mail:dphtemp@indiatimes.com

Printed at
Arora Offset Press
Laxmi Nagar, Delhi–92

Preface

This book Sampling and Attributes is written for graduate and post graduate students of all Indian universities. The subject matter of the book is presented in a very straight forward manner. So that the students are able to prepare this paper own and score well in the examination. Proof of various theorem and examples have been given with minute details.

Suggestions for the improvement of the book shall be grateful acknowledge and incorporated in the next edition.

The author is highly thankful to Mr. Tilak Wasan (Managing Director Discovery Publishing House) and Mr. Prem Pal Singh (Prop. Unique Computers) for bringing out this book in this present nice form.

A.K. Sharma

Contents

1

Sampling

INTRODUCTION

In our everyday life we come across persons making an assessment of the population through samples. Thus trader thrusts a conical trowel in a bag of wheat and assesses the quality of wheat in the bag by having a look upon the samples so drawn, a house-wife tests a small, quantity of rice to see if it has been well cooked and so on. The importance of the theory of sampling lies in the fact that for a large population, it is neither practical nor necessary to collect data for each and every member of the population. Thus in order to have an information about the economic conditions of the rural population of U.P., it would require a huge establishment to collect data from each and every individual in the villages and then to tabulate and calculate the parameters from it. It would be quite sufficient if a sample of village is selected (with due precautions) and informations collected from it. The information so gathered may be taken to represent the whole rural population of the state. Most of the industrial concerns have somesort of quality control over their manufactured goods before sending them to the market. If each and every item has to be tested, perhaps in some cases no goods can be sent to the market e.g., in the case of the match boxes, a test would mean the burning of all the matches. Thus the purpose of sampling is to get information about the population from the sample. The population measures e.g. mean, S,D. etc. are called parameters white the measures obtained from the samples are called *statistics*. The latter are *estimates* of their *paramters*.

Universe or Population may be defined as any collection of objects or results of an operation. Thus in the above example, the persons residing in villages of U.P. form the universe or population; similarly in some experiment, the data regarding pressures of a gas may form a population. A universe may contain a number of sub-universes e.g., the universe of studentsin a certain state may be divided into universe of students in primary schools, secondary schools, and colleges; similarly a universe may be part of other universe or universes e.g. the rural population of U.P. is a part of the whole population of U.P., which itself is a part of the population of India, then of Asia and so on.

By existent universe, we mean an aggregate of concrete objects such as the students in a college or the population of a city, while by *hypothetical universe* we mean all conceivable ways in which an event may happen e.g., all possible ways in which a coin tossed an indefinite number of times may fall; the number of ways in which the alphabet can be arranged to form a word.

In a finite universe, the number of objects is finite *e.g.*, the students in a college. In an infinite universe the number is finite *e.g.,* the pressures of a gas or the number of children born in a race.

Batch Size and Sample Size

Suppose a buyer agrees to buy goods in batches of 100 items each and out of each batch, he tests 5 items to check the quality. Then the *batch size* is 100 and 5 is the *sample size.* In general a sample with size more than 30 items is taken to be a large sample and it is independent of the batch size if the population is homogeneous. The tests in the case of large samples are not applicable in case of small samples.

Sampling is important in order to estimate the parameters of the population or to test certain hypothesis. It is a major tool for quality control

Census and Sample Method

Under the census or complete enumeration survey method, data are collected for each and every unit (person, household, field, shop, factory, etc., as the case may be) of the population or universe which is the complete set of items which are of interest in any particular situation.

(i) Data are obtained from each and every unit of the population.

(ii) The results obtained are likely to be more representative, accurate and reliable.

(iii) It is an appropriate method of obtaining information on rare events such as areas under some crops and yield thereof, the number of persons of certain age groups, their distribution by sex, educational level of people, etc. This is the reason why throughout the world the population data are obtained by conducting a census generally every 10 years by the census method.

(iv) Data of a complete enumeration census can be widely exploited as a basis for various surveys.

The process of sampling involves three elements:

(a) Selecting the sample,

(b) collecting the information, and

(c) Marking an inference about the population.

The three elements cannot generally be considered in isolation from one another. Sample selection, data collection, and estimation are all interwoven and each has an impact on the others. Sampling is not haphazard selection; it embodies definite rules for selecting, the sample. But having followed a set of rules for sample selection, we cannot consider the estimation process independent of it; estimation is guided by the manner in which the sample has been selected.

THEORETICAL BASIS OF SAMPLING

On the basis of sample study we can predict and generalise the behaviour of mass phenomena. This is possible because there is no statistical population whose elements would vary from each other without limit. For example, rice varies to limited extent in colour, protein content, length, weight etc., but it can always be identified as mango. Similarly apples of the same tree may vary in size, colour, taste, weight, etc. but they can always be identified as apples.

There are two important laws on which the theory of sampling is based:

1. Law of 'Statistical Regularity', and
2. Law of 'Inertia of Large Numbers'.

Law of Statistical Regularity

"The low of statistical regularity lays down that a moderately large number of items chosen at random from a large group are almost sure on the average to possess the characteristics of the large group." In other words, this law points out that if a sample is taken at random from a population, it is likely to possess almost the same characteristics as that of the population. According to this law desirability of choosing the sample at random.

Law of Inertia of Large Numbers

It states that, others things being equal, larger the size of the sample, more accurate the results are likely to be. This is because large numbers are more stable as compared to small ones. The difference in the aggregate result is likely to be insignificant, when the number in the sample is large, because when large numbers are considered the variation in the component part tend to balance each other and, therefore, the variation in the aggregate is insignificant.

Essentials of Sampling

If the sample results are to have any worthwhile meaning, it is necessary that a sample possesses the following essentials:

(i) *Representativeness* : A sample should be so selected that it truly represents the universe otherwise the results obtained may be misleading. To ensure representativeness the random method of selection should be used.

(ii) *Adequacy:* The size of sample should be adequate otherwise it may not represent the characteristics of the universe.

(iii) *Independence*: All items of the sample should be selected independently of one another and all items of the universe should have the same chance of being selected. in the sample. By independence of selection we mean that the selection of a particular item in one draw has no influence on the probabilities of selection in any other draw.

(iv) *Homogeneity:* When we talk of homogeneity we mean that there is no basic difference in the nature of units of the universe and that of the sample. If two samples from the same universe are taken, they should give more or less the same result .

METHODS OF SAMPLING

Types of Samples. The most important method of drawing a sample is that of random sampling, which implies that each member of the universe has the same chance of being selected for the sample. Great care has to be taken to ensure that the sample drawn is random. Generally selections made by human instinct, however carefully random, do contain some element of bias in it. In order to avoid individual bias, the selection is made with the help of random number tables, or some mechanical means like the drawing of a lottery, throw of a dice or reulette wheel etc.

Simple sampling is a particular case of random sampling in which the chance of selection of member of the universe for the sample is independent of the previous selection. Thus, if the selection is made with the help of a disc or toss of a coin, the sampling would be simple. However, if a pack of cards is to be used, the selection may not be simple. Suppose that we draw a sample of 10 out of 32 objects and to each card we assign a number corresponding to each member of the universe, the probability of individual members being selected in the successive draws would be 1/52, 1/51, 1/50... if the card drawn is not put back in the pack after each draw and though (he selection is random it is not simple. However, if after each draw the cards are replaced the sampling is simple. It may be noted that if the population is very large, the random sampling would always be simple.

The various methods of sampling or different sampling designs can be grouped under two broad heads-random sampling and non-random sampling. Random sampling is also referred to as probability sampling since if the sampling process is random the laws of probability can be applied. It may be noted that the term random sample is not used to describe the data in the sample but the process employed to select the sample. Randomness is thus a property of the sampling procedure instead of an individual sample. As such randomness can enter a processed sampling in a number of ways and hence random samples may be of many kinds.

Advantages of Probability Sampling

The following are the basic advantages of probability sampling :

(1) Probability sampling does not depend upon the existence of detailed information about the universe for its effectiveness.

(2) Probability sampling provides estimates which are essentially unbiased and have measurable precision.

(3) It is possible to evaluation the relative efficiency of various sample designs only when probability sampling is used.

Limitations of Probability Sampling

Despite the great advantages of probability sampling techniques mentioned above it has certain limitations because of which non-probability sampling is quite often used in practice. These limitations are:

(1) Probability sampling requires a very high level of skill and experience for its use.

(2) It requires a lot of time to plan and execute a probability sample.

(3) The costs involved in probability sampling are larger as compared to non-probability sampling.

Non-random sampling is a process of sample selection without the use of randomization. In other words, a non-random sample is selected on a basis other than the probability considerations such as convenience, judgement, etc.

The most important difference between random and non-random sampling is that whereas the pattern of sampling variability can be ascertained in case of random sampling, in non-random sampling, there is no way knowing the pattern of variability in the process.

Devices for Random Sampling

It has been experimentally established that selection by human instinct has always been subject to individual bias. Suppose a teacher is asked to select ten students from his college; he is more likely to select his favourite students. JB order to avoid this human factor, methods have been devised toseJect random samples. Some of these methods are given below.

Tippet's Numbers

L.H.C. Tippet constructed random number tables whicfl consist of 41600 digits taken from British census report to give 10400 four figure numbers. It has been found that those tables are fairly random and have played an important role in the sampling technique. The numbers may be taken according to the column, or row, or diagonally, on any page of the table.

Similarly, Fisher. Kendall and the Indian statistician Mahalnobis have published different random number tables. Random samples may be obtained by'the throw of a dice, draw of a lottery by mixing some numbered chits thoroughly and drawing lots by some mechanical means. This random figures are obtained by machines in the draw of *Lottery Prizes* in our country. RAND corporation of U.S.A. has prepared a random number table containing 1,000,000 random digits from random noise in an electrical circuits.

A. Random Sampling Methods

(a) Simple or unrestricted random sampling: and

(b) Restricted random sampling:

(i) Stratified sampling,

(ii) Systematic sampling,

(iii) Cluster sampling,

B. Non-Random Sampling Methods

(i) Judgment sampling;

(ii) Convenience sampling; and

(iii) Quota sampling.

A brief description of these methods is given below.

RANDOM SAMPLING METHODS

(a) Simple or Unrestricted Random Sampling

In simple random sampling which items get selected in the sample is just a matter of chance-personal bias of the investigator does not influence the selection. It should be noted that the word *'random'* does not mean *'haphazard'* or *'hit-or-miss'-it* rather means that the selection process is such that chance only determines which items are included in the sample. As pointed out by Chou, when a sample of size n is drawn from a population with N elements, the sample is a *'simple random sample'* if any of the following is true. And, if any of the following is true, so are the other two:

(1) All n items of the sample are selected independently of one another and all N items in the population have the same chance of being included in the sample. By independence of selection we mean that the selection of a particular item in one draw has no influence on the probabilities of selection in any other draw.

(2) At each selection, all remaining items in the population have the same chance of being drawn. If sampling is made with replacement *i.e.*, when each unit drawn from the population is returned prior to drawing the next unit, each item has a probability of 1/N of being drawn at each selection. If sampling is without replacement, *i.e.*, when each unit drawn from the population is not returned prior to drawing the next unit, the probability of selection of each item remaining in the population at the first draw is 1/N, at the second draw 1/(N–1), at the third draw is 1/(N–2), and so on. It should be noted that sampling with replacement has very limited and special uses in statistics-we are mostly concerned with sampling without replacement.

(3) All the possible samples of a given size n are equally likely to be selected.

To ensure randomness of selection one may adopt either the Lottery Method or consult table of random numbers.

Lottery Method

Under this method, all items of the universe are numbered or named on separate slips of paper of identical size and shape. These slips are then folded and mixed up in a container or drum. A blindfold selection is then made of the number of slips required to constitute the desired sample size. The selection of items thus depends entirely on chance. The methods would be quite clear with the help of an example. If we want to take a sample of 10 persons out of a population of 100, the procedure is to write the name of all the 100 persons on separate slips of paper, fold these slips, mix them thoroughly and then make a blindfold selection of 10 slips.

The random numbers are generally obtained by some mechanism which, when repeated a large number of times, ensures approximately equal frequencies for the numbers from 0 to 9 and also proper frequencies for various combinations of numbers (such as 00, 01,...99 ; 000,001,...999; etc.) that could be expected in a random sequence of the digits 0 to 9.

Several standard tables of random numbers are available, among which the following may be specially mentioned, as they have been tested extensively for randomness:

(1) Tippett's (1927) random number tables consisting of 41,600 random digits grouped into 10,400 sets of four-digited random numbers;

(2) Fisher and Yates (1938) table of random numbers with 15,000 random digits arranged into 1,500 sets of ten-digited random numbers.

(3) Kendall and B.B. Smith (1939) table of random numbers leaving 10,00,000 random digits grouped into 25,000 sets of four-digited random numbers;

(4) Rand corporation (1955) table of random numbers consisting of 1,00,000 random digits grouped into 2,00,000 sets of five-digited random numbers; and

(5) C.R. Rao, Mitra and Matthai (1966) table of random numbers with 20,000 random digits grouped into 5,000 sets of four-digited random numbers.

Merits

1. The analyst can easily assess the accuracy of this estimate because sampling errors follow the principle of chance. The theory of random sampling is further developed than that of any other type of sampling which enables the analyst to provide the most reliable information at the lest cost.
2. As compared to judgement sampling, a random sample represents the universe in a better way. As the size of the sample increases, it becomes increasingly representative of the population.
3. Since the selection of items in the sample depends entirely on chance there is no possibility of personal bias affecting the results.

Limitations

1. Random sampling may produce the most non-random-looking results. For example, thirteen cards from a well-shuffled pack of playing cards may consist of one suit. But the probability of this type of occurrence is very-very low.
2. From the point of view of field survey it has been claimed that case selected by random sampling tend to be too widely dispersed

geographically and that the time and cost of collecting data become too large.

3. The size of the sample required to ensure statistical reliability is usually larger under random sampling than stratified sampling.
4. The use of simple random sampling necessitates a completely catalogued universe from which draw the sample. But it is often difficult for the investigator to have up-to-date lists of all he items of the population to be sampled. This restricts the use of this method in economic and business date where very we have to employ restricted random sampling designs.

(b) Restricted Random Sampling

Stratified random sampling or simply stratified sampling is one of the random methods which, by using the available information concerning the population, attempts to design a more efficient sample than obtained by the simple random procedure.

Random sampling is not always the best method of assessing the population. Thus to estimate the average income of the inhabitants of a city, it is necessary that all sections of the society must be included in our sample otherwise there is a likelihood that more rich people or poor people may be dominating our sample. For this purpose, it would be better first to divide the city into different *Strata*, say, according to the localities slums, middle-class localities and bungalow areas, business localities etc., and then to select individuals at random from each of these localities. This would ensure that all sections of the society are represented in the sample. The above sampling technique is known as *stratified sampling.* The size of each group or starta should be proportionate to the relative importance in the population of the stratum represented by the group.

If the sampling is done according to the rules of probability the errors that are likely to creep in can be estimated and here lies the importance of random sampling. It is in this method that the rules of probability are applicable so that the statistics of the sample may be used to estimate the parameters of the populations, the results of the two samples may be compared.to test whether they have been drawn from the same universe and whether a hypothesis is to be rejected on the basis of the results of a sample. The sampling errors can always be reduced by increasing the sample size: it means corresponding increase in cost and labour.

When this method of sampling is adopted, the population is divided into different groups or classes called stratas and a sample is drawn from each stratum at random. For example, if we are interested in studying the consumption pattern of the people of Delhi, the city of Delhi may be divided into various parts (such as zones or wards) and from each part a sample may be taken at random. Before deciding on stratification we must have knowledge of the traits of the population. Such knowledge may be based upon expert judgment, past data, preliminary observation from pilot studies, etc.

The purpose of stratification is to increase the efficiency of sampling by dividing a heterogeneous universe in such a way that

(i) there is as great homogeneity as possible within each stratum and

(ii) as marked a difference as possible between the strata.

Example 1:

You are given the following data of the number of lecturers, readers and professors in a University:

Length of service	Lecturers	Readers	Professors	Total
Less than 5 Yrs.	2,000	250	50	2,300
5-10 Yrs.	3,000	220	80	3,300
10-15 Yrs.	1,500	170	30	1,700
More than 15 Yrs.	880	80	40	1,000
Total	7,380	720	200	8,300

Work out how many lecturers, readers and professors would be selected from each category if,

(i) we follow stratified proportionate sampling method and take 10% of the universe equivalent to the sample size,

(ii) if the size of the sample is 10% of the universe but the lecturers, readers and professors are to be in the ratio of 5:3:2 and weightage to the length of service is to be in the ratio of 4:3:2:1.

Solution :

(i) The sample size is 10% of the universe hence 830 persons would be selected in the sample. Since 12 strata are formed and we want to follow proportionate stratified sampling method. We will take 10% from each stratum. The number of Persons selected shall be as follows:

Length of service	Lecturers	Readers	Professors	Total
Less than 5 yrs.	200	25	5	230
5-10 yrs.	300	22	8	330
10-15 yrs.	150	17	3	170
Above 15 yrs.	88	8	4	100
Total	738	72	20	830

(ii) In the second case also the size of sample is 830 but the lecturers, readers and professors are to be in the ratio of 5:3:2 of the sample, *i.e.*, we take $\frac{830 \times 5}{10} = 415$ lecturers, $\frac{830 \times 3}{10} = 249$ readers, and $\frac{830 \times 2}{10} = 166$ professors. Since the weightage to length of service is 4 : 3 : 2 : 1 the number selected from each category shall be :

Length of service	Lecturers	Readers	Professors	Total
Less than 5 yrs.	$\frac{415 \times 4}{10} = 166$	$\frac{249 \times 4}{10} = 99.6$ or 100	$\frac{166 \times 4}{10} = 66.4$ or 66	332
5 – 10	$\frac{415 \times 3}{10} = 124.5$ or 124	$\frac{249 \times 3}{10} = 74.7$ or 74	$\frac{166 \times 3}{10} = 49.80$ or 50	248
10 – 15	$\frac{415 \times 2}{10} = 83$	$\frac{249 \times 2}{10} = 49.8$ or 50	$\frac{166 \times 2}{10} = 33.2$ or 33	166
Above 15	$\frac{415 \times 1}{10} = 41.5$ or 42	$\frac{249 \times 1}{10} = 24.9$ or 25	$\frac{166 \times 1}{10} = 16.6$ or 17	84
Total	415	249	166	830

Merits

1. Greater geographical concentration. As compared with random sample, stratified samples can be more concentrated geographically *i.e.*, the units from the different strata may be selected in such a way the all of them are localised in one geographical area. This would greatly reduce the time and expenses of interviewing.
2. Greater accuracy. Stratified sampling ensures greater accuracy. The accuracy is maximum if each stratum is so formed that it consists of uniform or homogeneous items.
3. More representative. Since the population is first divided into various strata and then a sample is drawn from each stratum there is little possibility of any essential group of the population being completely excluded. A more representative sample is thus secured. C.J. Grohmann has rightly pointed out that this type of sampling balances the uncertainity of random sampling against the bias of deliberate selection."

Limitations

1. The items from each stratum should be selected at random. But this may be difficult to achieve in the absence of skilled sampling supervisors and a random selection within each stratum may not be ensured.
2. Utmost care must be exercised in dividing the population into various stratas. Each stratum must contain, as possible, homogeneous items as otherwise the results may not be reliable. If proper stratification of the population is not done the sample may have the effect of bias.

(ii) Systematic Sampling

A systematic sample is formed by selecting one unit at random and then selecting additional units at evently-spaced intervals until the sample has been formed. ; This method is popularly used in those cases where a complete list of the population from which sample is to be drawn is available. The list may be prepared in alphabetical, geographical, numerical or some other. The items are serially numbered. The first item is selected at random generally by following the Lottery method. Subsequent items are selected are selected by taking every kth item from the list where 'k' refers to the sampling interval or sampling ratio *i.e.*, the ratio of population size of the sample. Symbolically,

$$k = \frac{N}{n}$$

where k = Sampling interval, N = Universe size and n = Sample size.

While calculating k, it is possible that we get a fractional value. In such a case we should use approximation procedure *i.e.*, if the fraction is less than 0.5 it should be omitted and if it is more than 0.5 it should be taken as 1. If it is exactly 0.5 it should be omitted if the number is even and should be taken as 1, if the number is odd. This is based on the principle that the number after approximation should preferably be even. For example if the number of students are respectively 1020, 1150 and 1100 and we want to take a sample of 200, k shall be:

$$\text{(i) } 8\ k = \frac{1020}{200} = 5.1 \text{ or } 5$$

$$\text{(ii) } 82\ k = \frac{1150}{200} = 5.75 \text{ or } 6$$

$$\text{(iii) } 2\ k = \frac{1100}{200} = 5.5 \text{ or } 6\ s$$

Example 2 :

In a class there are 96 students with Roll Nos. from 1 to 96. It is desired to take a sample of 10 students. Use the systematic sampling method to

Solution :

$$k = \frac{N}{n} = \frac{96}{10} = 9.6 \text{ or } 10$$

From 1 to 96 Roll Nos. The first student between 1 and k *i.e.*, 1 and 10 will be selected at random and then we will go on taking every kth student. Suppose the first student comes out to be 4th. The sample would them consist of the following Roll Nos.

4, 14, 24, 434, 44, 54, 64, 74, 84, 94,

Systematic sampling is relatively a simple technique and may be more efficient than simple random sampling provided the lists are arranged wholly at random. However, it is rarely that this requirement is fulfilled. The nearest approach to randomness is provided by alphabetical lists such as are found in telephone directory, although even these may have certain non-random characteristics.

Merits

The systematic sampling design is simple and convenient to adopt. The time and work involved in sampling by this method are relatively smaller.

The result obtained are also found to be generally satisfactory provided care is taken to see that there are no periodic features associated with the sampling interval. If populations are sufficiently large systematic sampling can often be expected to yield results similar to those obtained by proportional stratified sampling.

Limitation

The main limitation of the method is that it becomes less representative if we are dealing with populations having hidden *periodicities*. Also if the population is ordered in a systematic way with respect to the characteristics the investigator is interested in, then it is possible that only certain types of items will be included in the population, or at least more of certain types than others. For instance, in a study of workers' wages the list may be such that every tenth worker on the list gets wages above Rs. 150 per month.

(iii) Multi-stage or Sampling Cluster Sampling

As the name implies this method refers to a sampling procedure which is carried out in several stages. The material is regarded as made up for a number of second stage sampling units, each of which is made of a number of second stage units, etc. At first, the first stage units are sampled by some suitable method, such as simple random sampling, Then, a sample of second stage units is selected from each of the selected first stage units, again by some suitable method which may be the same as or different from the method employed for the first stage units. Further stages may be added as required. The producer may be illustrated as follows:

Suppose we want to take a sample of 5,000 households from the Stage of U.P. At the first stage, the Stage may be divided into a number of districts and a few districts selected at random. At the second stage, each district may be sub-divided into a number of villages and a sample of villages may be taken at random. At the third stage, a number of households may be selected from each of the villages selected at the second stage. To take another example suppose in a particular survey, we wish to take a sample of 10,000 students from Delhi University. We may take colleges primary units-at the first stage, then draw departments as the second stage, and choose students at the third and last stage.

Merits

Multi-stage sampling introduces flexibility in the sampling method which is lacking in the other methods. It enables existing divisions and sub-divisions of the population to be used as units at various stages, and permits the field work to be concentrated and yet large area to be covered, Another advantage of the method is that sub-division into second stage units (*i.e.*, the construction of the second stage frame) need be carried out for only those first stage units which are included in the sample. It is, therefore, particularly valuable in surveys of underdeveloped area where no frame is generally sufficiently detailed and accurate for sub-division of the material into reasonably small sampling units.

Limitations

However, a multi-stage sample is in general less accurate than a sample containing the same number of final stage units which have been selected by some suitable single stage process.

We have discussed above the various random procedures as independent design. In practice we often combine two or more of these methods into a single design.

B. Non-Random Sampling Methods

(i) Judgment Sampling

In judgment sampling the choice of sampling items depends exclusively on the discretion of the investigator. In other words, the investigator exercises his judgment in the choice and includes those items in the sample which he thinks are most typical of the universe with regard to the characteristics under investigation. For example, if sample of ten students is to be selected from a class of sixty for analysing the spending habits of students, the investigator would select 10 students who, in his opinion, are representative of the class.

Merits

Though the principles of sampling theory are not applicable to judgement sampling, the method is often used in solving many types of economic and business problems. The use of judgment sampling is justified under a variety of circumstance:

(i) In solving everyday business problems and making public policy decisions, executives and public officials are often pressed for time and cannot wait for probably sample designs. Judgment sampling is then the only practical method to arrive at solutions to their urgent problems.

(ii) When we want to study some unknown traits of a population, some of whose characteristics are known, we may then stratify the population according to these known properties and select sampling units from each stratum on the basis of judgment. This method is used to obtain a more representative sample.

(iii) when only a small number of sampling units is in the universe, simple random selection may miss the more important elements, whereas judgment selection would certainly include them in the sample.

Limitations

This method, though simple, is not scientific because the population units to be sampled may be affected by the personal prejudicer or bias of the investigator. Thus, judgement sampling involves the risk that the investigator may establish foregone conclusions by including those items in the sample which conform to his preconceived notions. For example, if an investigator holds the view that the wages of workers in a certain establishment are very low, and if he adopts the judgment sampling method, he may include only those workers in the sample whose wages are low and thereby establish his point of view which may be far from the truth. Since an element of subjectiveness is possible, this method cannot be recommended for general use.

(ii) Convenience Sampling

A convenience sample is obtained by selecting 'convenient' population units.

The method of convenience sampling is also called the chunk. A chunk refers to that fraction of the population being investigated which is selected neither my probability nor by judgment but by convenience. A sample obtained from readily available lists such as automobile registrations, telephone directories, etc. is a convenience sample and not a random sample even if the sample is drawn at random from the lists. If a person is to submit a project report on labour-management relations in textile industry and he takes a textile mill close to his office and interview some people over there, he is following convenience sampling method. Convenance samples are prove to bias by their very nature-selecting population elements which are convenient to choose almost always makes them special or different from the best of the elements in the population is some way.

(iii) Quota Sampling

Quota sampling is a type of judgment sampling. In a quota sample, quotas are set up according to some specified characteristics such as so many in each of several income groups, so many in each age, so many with certain political or religious affiliations, and so on. Each interviewer is then told to interview a certain number of persons which constitutes his quota. Within the quotas, the selection of sample items depends on personal judgment.

Quota sampling is often used in public opinion studies. It occasionally provides satisfactory results if the interviewers are carefully trained and if they follow their instructions closely.

Selection of Appropriate Method of Sampling

Having discussed the various methods of sampling, the question now arises as to which method to adopt in a particular situation. It should be noted that one method can be regarded as best under all circumstances- each method has its own speciality. A number of factors such as nature of the problem, size of universe, size of the sample, availability of finance time, etc. would influence the selection of a particular method of sampling.

SIZE OF SAMPLE

An important decision that has to be taken adopting a sampling technique is about the size of the sample. Size of sample means the number of sampling units selected from the population for investigations. Different opinions have been expressed by experts on this point.

The following factors should be considered while deciding the sample size:

(i) *Nature of Study:* For an intensive and continuous study a small sample may be suitable. But for studies which are not likely to be repeated and are quite extensive in nature, it may be necessary to take a large sample size.

(ii) *Homogeneity or Heterogeneity of the Universe* : If the universe consists of homogeneous units a small sample may be inevitable.

(iii) *The degree of accuracy or precision desired* : The greater the degree of accuracy desired the larger should be the sample always ensure greater accuracy. If a sample is selected by experts by following scientific method, it may ensure better results even when it is small compared to a situation in which a large sample size is selected by inexperienced people.

(iv) *The resources available* : If the resources available are vast a larger sample size could be taken. However, in most cases resources constitute a big constraint on sample size.

(v) *The size of the universe:* The larger the size of the universe, the bigger should be the sample size.

(iv) *Nature of respondents:* Where it is expected that a large number of respondents, will not cooperate and send back the questionnaires, a larger sample should be selected.

The above factors have to be properly weighed before arriving at the sample size. However, the selection of optimum sample size is not that simple as it might seem to be. If the sample is used which is larger than necessary, resources are wasted, if the sample is smaller than required, the objectives of the analysis man not be achieved.

(vii) *Methods of Sampling adopted:* The size of sample is also influenced by the type of sampling plan adopted. For example, if the sample is a simple random sample it may necessitate a bigger sample size. However, in a properly drawn stratified sampling plan, even a small sample may give better results.

Mathematical Formula for Determining the Sample Size

A number of formulae have been devised for determining the sample size depending upon the availability of information. A few formulae are given below :

$$n = \left(\frac{Z\sigma}{d}\right)^2$$

n = Sample size

Z = value at a specified level of confidence or desired degree of precision.

s = Standard deviation of the population

d = difference between population mean and sample mean.

The steps in computing the sample from the above formula are:

(i) Select the desired degree of precision, *i.e.*, specified level or confidence and designate it as small 'z' (at 1% level of significance or 99% confidence level the valde of 'z' is 2.576, and at 5% level of significance of 95% confidence level 1.96).

(ii) Multiply the 'z' selected in step 1 by the standard deviation of the universe which may be assumed.

(iii) Divide the product of the preceding step by the standard error of mean or difference between population and sample mean. Square the resultant quotient. The result is the size of sample required.

MERITS AND LIMITATIONS OF SAMPLING

Merits

The sampling technique has the following merits over the complete enumeration survey:

1. *More reliable results :* Although the sampling technique involves certain inaccuracies owing to sampling errors, the result obtained is generally more reliable than that obtained from a complete count. There are several reasons for it. First, it is always possible to determine the extent of sampling errors. Secondly, other types of errors to which a survey is subject, such as inaccuracy of information, incompleteness of returns, etc., are likely to be more serious in a complete census than in a sample survey. This is because more effective precautions can be taken in a sample survey to ensure that the information is accurate and complete. For these reasons not only may the total error be expected to the smaller in a sample survey but sample results can also be used with a greater degree of confidence because of our knowledge of the probable size of error. Thirdly, it is possible to avail of the services of experts and to impart thorough training to the investigators in a sample survey which further reduces the possibility of errors. Follow-up work can also be undertaken much more effectively in the sampling method. Indeed, even a complete census can only be tested for accuracy by some type of sampling check.
2. *Less time consuming :* Since the sample is a study of a part of population considerable time and labour are saved when a sample survey is carried out. Time saved not only in collecting data but also in processing it. For these reasons a sample provides more timely data in practice that a census.
3. *Less cost :* Although the amount of effort and expense involved in collecting information is always greater per unit of the sample than a complete census, the total financial burden of a sample survey is generally less than that of a complete census. This is because of the fact that in sampling, we study only a part of population and the total expenses of collecting data is less than that required when the census method is adopted. This is a great advantage particularly in an underdeveloped economy where much of the information would be difficult to collect by the census method for lack of adequate resources.

4. *More detailed information :* Since the sampling technique saves time and money, it is possible to collect more detailed information in a sample survey. For example, if the population consists of 1,000 persons in a survey of the consumption pattern of the people, the two alternative techniques available are as follows: .
   (a) We may collect the necessary data from each one of the 1,000 people through a questionnaire containing, say, 10 questions (census method), or
   (b) We may take a sample of 100 persons (*i.e.*, 10% of population and prepare a questionnaire containing as many as 10 questions. The expenses involved in the latter case would almost be the same as in the former but it will enable nine time more information to be obtained.
5. The sample method is often used to judge the accuracy of the information obtained on a census basis. For example, in the population census which is conducted very often 10 years in our country the field officers employ the sample method to determine the accuracy of information obtained by the enumerators on the census basis.
6. *Sampling Method is the only method that can be used in certain cases.* There are some cases in which the census method is inapplicable and the only practicable means is provided by the sample method. For example, if one is interested in testing the breaking strength of chalks manufactured in a factory under the census method all the chalks would be broken in the process of testing. Hence, census method is impracticable and resort must be had to the sample method. Similarly if the producer wants to find out whether the tensile strength of a lot of steel wires meets the specified standard. He must resort to sample method because census would mean complete destruction of all the wires. Also if the population under investigation is infinite, sampling is the only possible solution.

Limitations

Despite the various advantages of sampling, it is not altogether free from limitations. Some of the difficulties involved in sampling are stated below:

1. If the information is required for each and every unit in the domain of study, a complete enumeration survey is necessary.
2. At time the sampling plan may be so complicated that it requires more time, labour and money than a complete count. This is so if the size of the sample is a large proportion of the total population

and if complicated weighed procedures are used. With each additional complication in the survey, the chance of errors multiply and greater care has to be taken which, in turn, means more time and labour.

3. Sampling generally requires the services of experts, if only for consultation purposes. In the absence of qualified and experienced persons, the information obtained from sample surveys cannot be relied upon. In India, shortage of experts in the sampling field is a serious hurdle in the way of reliable statistics.
4. A sample survey must be carefully planed and executed otherwise the results obtained may be inaccurate and misleading. Of course, even for a complete count care must be taken but serious errors may arise in sampling, if the sampling procedure is not perfect.

SAMPLING AND NON-SAMPLING ERRORS

To appreciate the need for sample surveys, it is necessary to understand clearly the role of sampling and non-sampling errors in complete enumeration and sample surveys. The error arising due to drawing inferences about the population on the basis of few observations (sampling) is termed *sampling error.* Clearly the sampling error in this sense is nonexistent in complete enumeration survey, since the whole population is surveyed. However, the error mainly arising at the stages of ascertainment and processing of data, which are termed non-sampling errors, are common both in complete enumeration and sample surveys.

I. Sampling Errors

Even if utmost care has been taken in selecting a sample. The results derived from a sample study may not be exactly equal to the true value in the population. The reason is that estimate is based on a part and not on the whole and samples are seldom, if ever, perfect miniature of the population. Hence sampling gives rise to certain errors known as sampling errors (or sampling fluctuations). These errors would not be present in a complete enumeration survey. However, these errors can be controlled. The modern sampling theory helps in designing the survey in such a manner that the sampling errors can be made small.

Sampling errors are of two types-biased and unbiased.

(1) *Biased errors.* These errors arise from any bias in selection, estimation, etc.

(2) *Unbiased errors.* These errors arise due to chance differences, between the members of population included in the sample and those not included. An error in statistics is the difference between the value of a statistic and that of the corresponding parameter.

Thus, the total sampling error is made up of error due to bias, if any, and the random sampling error. The essence of bias is that it forms a constant component of error that does not decrease in a large population as the number in the sample increases. Such error is, therefore, also known as *cumulative or non-compensating error.* The random sampling error, on the other hand, decreases on an average as the size of the sample increases. Such error is, therefore, also known as non-cumulative or compensating error.

Causes of Bias

Bias may arise due to:

(i) faulty process of selection;

(ii) faulty work during the collection of information; and

(iii) faulty methods of analysis.

(i) Faulty selection : Faulty selection of the sample may give rise to bias in a number of ways, such as:

(1) An appeal to the vanity of the person questioned may give rise to yet another kind of bias. For example, the question 'Are you a good student?' Is such that most of the students would succumb to vanity and answer 'Yes.'

(2) *Non-response.* If all the items to be included in the sample are not covered, there will be bias even though no substitution has been attempted. This fault particularly occurs in mailed questionnaires, which are incompletely returned. Moreover, the information supplied by the informants may also be biased.

(3) *Substitution.* Substitution of an item in place of one chosen in a random sample sometimes leads to bias. Thus if it was decided to interview every 50th householder in the street, it would be inappropriate to interview the 51st or any other number in his place as the characteristics possessed by them may differ from those who were originally to be included in the sample.

(4) *Conscious or unconscious bias in the selection of a 'random' sample.* The randomness of selection may not really exist, even though the investigator claims that he has a random sample if he allows his desire to obtain a certain result to influence his selection.

(5) Deliberate selection of a 'representative' sample.

(ii) Bias due to Faulty Collection of Data : Any consistent error in measurement will give rise to bias whether the measurements are carried out on a sample or on all the units of the population. The danger of error is, however, likely to be greater in sampling work, since the units measured are often smaller. Bias may arise due to improper formulation of the decision, problem wrongly defining the population, specifying the wrong decision, securing an inadequate frame, and so on. Biased observations may result from a poorly designed questionnaire, an ill-trained interviewer, failure of a respondent's memory, etc. Bias in the flow of data may be due to unorganised collection procedure, faulty editing or coding of responses.

(iii) **Bias in Analysis :** In addition to bias which arises from faulty process of selection and faulty collection of information, faulty methods of analysis may also introduce bias. Such bias can be avoided by adopting the proper methods of analysis.

Avoidance of Bias

If possibilities of bias exist, fully objective conclusions cannot be drawn. The first essential of any sampling or census procedure must, therefore, be the elimination of all sources of bias. The simplest and the only certain way of avoiding bias in the selection process is for the sample to be drawn either entirely at random, or at random, subject to restrictions which, while improving the accuracy, are of such a nature that they do not introduce bias in the results. In certain cases, systematic selection may also be permissible.

METHOD OF REDUCING SAMPLING ERRORS

Once the absence of bias has been ensured, attention should be given to the random sampling errors. Such errors must be reduced to the minimum so as to attain the desired accuracy.

Apart from reducing errors of bias, the simplest way of increasing the accuracy of a sample is to increase its size. The sampling error usually decreases with increase in sampling size (number units selected in the samples) and in act in many situations the decrease is inversely proportional to the square-root of the sample size as can be seen from the diagram below.

From this diagram it is clear that though the reduction in sampling error is substantial for initial increases in sample size, it becomes marginal after a certain stage..

II. Non Sampling Errors

When a complete enumeration of units in the universe is made one would expect that it would give rise to data free from errors. However, in practice it is not so.

Non-sampling errors can occur at every stage of planning and execution of the census or survey. Such errors can arise due to a number of causes such as defective methods of data collection and tabulation, faulty definition, incomplete coverage of the population or sample, etc. More specifically, non-sampling errors, may arise from one or more of the following factors:

1. Errors committed during presentation and printing of tabulated results.
2. Errors in data proceeding operations such as coding, punching, verification, etc.
3. Errors due to non-response, *i.e.*, incomplete coverage in respect of units.
4. Lack of adequate inspection and supervision of primary staff.
5. Lack of trained and experienced investigators.
6. Inaccurate or inappropriate methods of interview, observation or measurement with inadequate or ambiguous schedules, definitions or instructions.
7. Inappropriate statistical unit.
8. Data specification being inadequate and inconsistent with respect to the objectives of the census or survey.

These sources are not exhaustive, but are given to indicate some of the possible sources of error. In a sample survey, non-sampling errors may also arise due to defective frame and faulty selection of sampling units.

Control of Non-sampling Errors : In some situations the non-sampling errors may be large and deserve greater attention than the sampling errors. While in general, sampling errors decreases with increase in sample size, non-sampling errors tend to increase with the sample size. In the case of complete enumeration non-sampling error and in the case of sample surveys both sampling and non-sampling errors require to be controlled and reduced to a level at which their presence does not vitiate the use of final results.

How to judge the Reliability of Samples

The reliability of samples can be tested in the following ways:

(1) Sub-samples should be taken from the sample and studied. If the results of sample and sub-sample study show similarity, the sample will be reliable.

(2) If the measurements of the universe are known then they should be compared with the measurements of the sample. In case of similarity of measurements, the sample will be reliable.

(3) More samples of the sample size should be taken from the same universe and their results be compared. If results are similar, the sample will be reliable.

2

Skewness, Moments and Kurtosis

INTRODUCTION

There are two other comparable characteristics called skewness and kurtosis that help us to understand a distribution. Two distributions may have the same mean and standard deviation but may differ widely in their overall appearance as can be seen from the following:

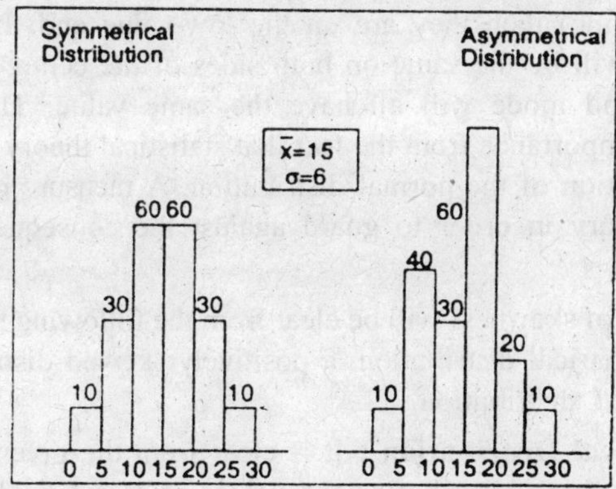

In both these distributions the value of mean and standared deviation is the same ($\overline{X}$ = 15, σ = 5). But it does not imply that the distributions are alike in nature. The distribution on the left-hand side is a symmetrical one whereas the distribution on the right-hand side is symmetrical or skewed. Measures of skewness help us to distinguish between different types of distributions.

Some important definitions of skewness are given as follows:

1. "A distribution is said to be 'skewed' when the mean and the median fall at different points in the distribution, and the balance (or centre of gravity) is shifted to one side or the other–to left or right."

—By-Garrett

2. "Measures of skewness tell us the direction and the extent of skewness. In symmetrical distribution the mean, median and mode are identical. The more the mean moves away from the mode, the larger the asymmetry or skewness". *—By Simpson & Kafka*
3. "Skewness refers to the asymmetry or lack of symmetry in the shape of a frequency distribution." *—By Morris Hamburg*
4. "When a series is not symmetrical it is said to be asymmetrical or skewed." *—By Croxtion & Cowden*

The analysis of above definitions shows that the term 'skewness' refers to tack of symmetry, *i.e.*, when a distribution is not symmetrical (or is asymmetrical) it is called a skewed distribution. Any measure of skewness indicates the difference between the manner in which items are distributed in a particular distribution compared with a symmetical (or normal) distribution. If, for example, skewness is positive, the frequencies in the distribution are spread out over a greater range of values on the high-value end of the curve (the right-hand side) then they are on the low-value end. If the curve is normal. Spread will be the same on both sides of the centre point and the mean, median and mode will all have the same value. The concept of skewness gains importance from the fact that statistical theory is often based upon the assumption of the normal distribution. A measure of skewness is therefore, necessary in order to guard against the consequences of this assumption.

The concept of skewness will be clear from the following three diagrams showing a symmetrical distribution, a positively skewed distribution and a negatively skewed distribution.

1. **Symmetrical Distribution :** It is clear from the given diagram that in a symmetrical distribution the values of mean, median and mode coincide. The spread of the frequencies is the same on both sides of the centre point of the curve.

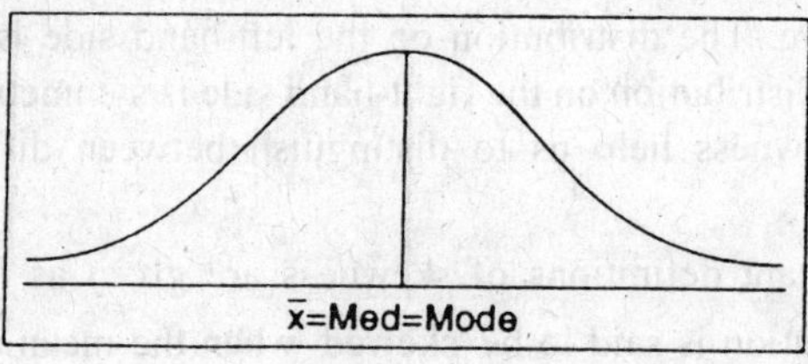

2. **Asymmetrical Distribution.** A distribution which is not symmetrical is called a skewed distribution and such a distribution could either be positively skewed or negatively skewed it is clear from the following diagram:

3. **Positively Skewed Distribution.** In the positively skewed distribution. The value of the mean is maximum and that of mode least–the median lies in between the two, it is clear from the following diagram:

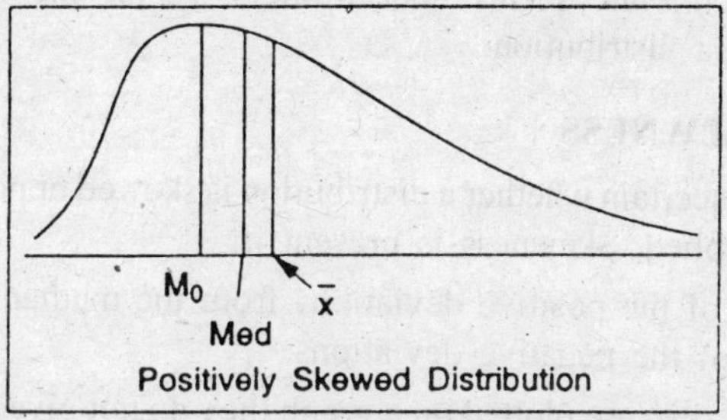

Positively Skewed Distribution

4. **Negatively Skewed Distribution.** The shape os negatively skewed distribution is as follows:

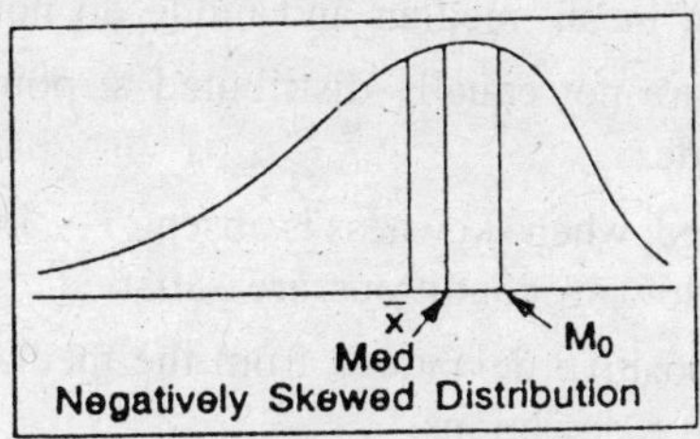

Negatively Skewed Distribution

In a negatively skewed distribution the value of mode is maximum and that of mean least–the median lies in between the two. In the positively skewed distribution the frequencies are spread out over a greater range of values on the high-value end of the curve (the right-hand side) than they are on the low-value end. In the negatively skewed distribution the position is reversed, *i.e.*, the excess tail is on the left-hand side

It should be noted that in moderately symmetrical distributions the interval between the mean and the median is approximately one-third of the interval between the mean and the mode. It is this relationship which provides a means of measuring the degree of skewness.

Difference Between Dispersion and Skewness

Dispersion is concerned with the *amount of uariation* rather than with its *direction.* Skewness tells us about the direction of the variation or the departure from symmetry. In fact, measures of skewness are dependent upon the amount of dispersion.

It may be noted that although skewness is an important characteristic for refining the precise pattern of a distribution, *it is rarely calculated in business and economic series.* Variation is by far the most important characteristic of a distribution.

TESTS OF SKEWNESS

In order to ascertain whether a distribution is skewed or not, the following tests may be applied. Skewness is present if:

(a) The sum of the positive deviations from the median is not equal to the sum of the negative deviations.

(b) When the data are plotted on a graph they do not give the normal bell-shaped form, *i.e.*, when cut along a vertical line through the centre the two halves are not equal.

(c) Quartiles are not equidistant from the median.

(d) The values of mean, median and mode do not coincide.

(e) Frequencies are not equally distributed at points of equal deviation from the mode.

Conversely stated, when skewness is absent, *i.e.*, in case of a symmetrical distribution, the following conditions are satisfied:

(a) Sum of the positive deviations from the median is equal to the sum of the negative deviations.

(b) Data when plotted on a graph give the normal bell-shaped form.

(c) Quartiles are equidistant from the median.

(d) The values of mean, median and mode coincide.

(e) Frequencies are equally distributed at points of equal deviations from the mode.

MEASURES OF SKEWNESS

Measures of skewness tell us the direction and extent of asymmetry in a series and permit us to compare two or more series with regard to these. They may either be absolute or relative.

Absolute Measures of Skewness

Skewness can be measured in absolute terms by taking the difference between mean and mode. Symbolically.

$$\text{Absolute Sk}^* = \overline{X} - \text{Mode.}$$

If the value of mean is greater than mode skewness will be positive, i.e, we shall get a plus sign in the above formula. Conversely, if the value of mode is greater than mean, we shall get a minus sign meaning thereby that the distribution is negatively skewed.

The reason why the difference between mean and mode can be used to measure skewness is that in a symmetrical distribution the values of mean, median and mode are alike, but the mean moves away from the mode when the observations are asymmetrical. Consequently, the distance between the mean and the mode could be used to measure skewness–the grater this distance, whether positive or negative, the more asymmetrical the distribution. However, such a measure is unsatisfactory on two counts:

1. If would be expressed in the unit of value of the distribution and could, therefore, not be compared with another comparable series expressed in different units.
2. Distributions vary greatly and the difference between, say, the Mean and the Mode in absolute terms might be considerable in one series and small in another, although the frequency curves of two distributions were similarly skewed.

If the absolute differences were expressed in relation to some measure of the spread of values in their respective distributions, the measure would be relative and can be directly for comparison. This leads us to the discussion of the relative measures of skewness.

Relative Measures of Skewness

There are four important measures of relative skewness, namely.

(a) The Karl Pearson's coefficient of skewness.

(b) The Bowley's coefficient of skewness.

(c) The kelly's coefficient of skewness.

(d) Measure of skewness based on moments.

These measures of skewness should mainly be used for making comparison between two or more distributions. As a description of one distribution alone, the interpretation of a measure of skewness is necessarily vague as "slight skewness", "marked skewness", or "moderate skewness".

A good measure of skewness should have the following three properties. It should:

1. Have a zero value, when the distribution is symmetrical; and
2. Have some meaningful scale of measure so that we could easily interpret the measured value.
3. Be a pure number in the sense that its value should be independent of the units of the series and also of the degree of variation in the series;

1. Karl Pearson's Coefficient of Skewness

This method of measuring skewness, also known as Pearsonisn Coefficient of Skewness, was suggested by Karl Pearson*, a great British Biometrician and Statiscian. It is based upon the difference between mean and mode. This difference is divided by standard deviation to give a relative measure. The formula thus becomes:

$$Sk_P = \frac{\text{Mean} - \text{Mode}}{\text{S tan dared Deviation}} \qquad \text{(i)}$$

Sk_P = Karl Pearson's coefficient of skewness.

There is no limit to this measure in theory and this is a slight drawback. But in practice the value given by this formula is rarely very high and usually lies between ± 1.

When a distribution is symmetrical, the values of mean, median and mode coincide and, therefore, the coefficient of skewness will be zero. When a distribution is positively skewed, the coefficient of skewness shall have plus sign and when it is negatively skewed, the coefficient of skewness shall have minus sing. The degree of skewness shall be obtained by the numbertial value, say, 0.8 or 0.2 etc. Thus, this formula gives both the *direction* as well as the *extent of skewness.*

The above method of measuring skewness cannot be used where mode is ill defined. However, in moderately skewed distribution the avarages have the following relationship:

$$\text{Mode} = 3 \text{ Median} - 2 \text{ Mean}$$

and therefore, if this value of mode is substituted in the above formula we arrive at another formula for finding out skewness.

$$Sk_P = \frac{[\overline{X} - (3 \text{ Med.} - 2\overline{X}]}{\sigma} = \frac{\overline{X} - 3 \text{ Med.} + 2\overline{X}}{\sigma} = \frac{3(\overline{X} - \text{Med.})}{\sigma} \quad \text{...(ii)}$$

Theoretically, the value of this coefficient varies between ± 3; however, in practice it is rare that the coefficient of skewness obtained by the above method exceeds ± 1.

2. Bowley's Coefficient of Skewness

An alternative measure of skewness has bee proposed by the late professor Bowley. Bowley's measure is based on *quartiles.* In a symmetrical distribution first and third quartiles are equidistant from the median as can be seen from the following diagram.

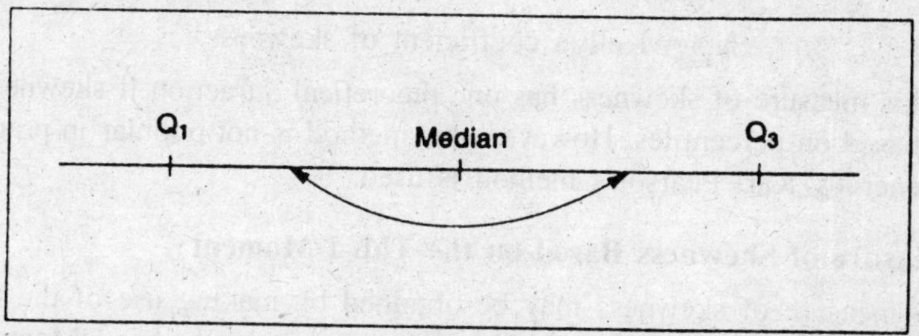

In a asymmetrical distribution the third quartile is the same distance over the median as the first quartile is below it, *i.e.*,

$$Q_3 - \text{Med.} = \text{Med.} - Q_1 \text{ or } Q_3 + Q_1 - 2 \text{ Med.} = 0$$

If this distribution is positively skewed the top 25 per cent of the values will tend to the farther from median than the bottom 25 per cent, *i.e.*, Q_3 will be farther from median than Q_1 is from median and the reverse for negative skewness. Hence a possible measure is

$$Sk_B = \frac{(Q_3 - \text{Med.}) - (\text{Med.} - Q_1)}{(Q_3 - \text{Med.}) + (\text{Med.} - Q_1)} \Rightarrow \frac{Q_3 + Q_1 - 2 \text{ Med.}}{Q_3 - Q_1}$$

Sk_B = Bowley's coefficient of skewness.

It must be remembered that the results obtained by these two measures are not to be compared with one another. Especially, the numberical values are not related to one another since the Bowley's measure, because of its computational basis, is limited to values between – 1 and + 1, while Pearson's measure has no such limits.

Not only do the numberical values obtained from these two formulae bear no necessary relationship to one another but, on rare occasions, with usually shaped distributions, its is possible for them to emerge with opposite signs.

3. Kelly's Coefficient of Skewness

Bowley's measure discussed above neglects the two extreme quarters of the data. It would be better for a measure to cover the entire data especially because in measuring skewness, we are often interested in the more extreme items. Bowley's measure can be extended by taking any two decides equidistant from the median or any two percentiles equidistant from the median. Kelly has suggested the following formula for measuring skewness upon the 10th and the 90th percentiles (or the first and ninth deciles):

$$Sk_K = \frac{P_{10} + P_{90} - 2\text{ Med.}}{P_{90} - P_{10}} \text{ also } Sk = \frac{D_1 + D_9 - 2\text{ Med.}}{D_9 - D_1}$$

Sk_K = Kelly's coefficient of skewness.

This measure of skewness has one theoretical attraction if skewness is to be based on percentiles. However, this method is not popular in practice and generally Karl Pearson's method is used.

4. Measure of Skewness Based on the Third Moment

A measure of skewness may be obtained by making use of the third moment about the mean. This would be discussed under the head 'Moments'.

The following illustrations shall explain the application of the above methods.

Example 1:

Calculate Karl Pearson's coefficient of skewness from the following data:

Profits (Rs. Lakhs)	*No. of Cos.*	*Profits (Rs. Lakhs)*	*No. of Cos.*
70-80	*12*	*110-120*	*50*
80-90	*18*	*120-130*	*45*
90-100	*35*	*130-140*	*30*
100-110	*42*	*140-150*	*8*

Solution:

Calculation of Coeff. of Skewness by Karl Pearson's Method

Profits (Rs. Lakhs)	**m.p. m**	**f**	**(m–115)/10 d**	**fd**	**fd²**
70-80	75	12	–4	–48	192
80-90	85	18	–3	–54	162
90-100	95	35	–2	–70	140
100-110	105	42	–1	–42	42
110-120	115	50	0	0	0
120-130	125	45	+1	+45	45
130-140	135	30	+2	+60	120
140-150	145	8	+3	+24	72
		N = 240		$\Sigma fd = -85$	$\Sigma fd^2 = 773$

$$\text{Coeff. of Sk.} = \frac{\text{Mean} - \text{Mode}}{\sigma}$$

$$\text{Mean:} \quad \overline{X} = A + \frac{\Sigma fd}{N} \times i = 115 - \frac{85}{240} \times 10$$

$$= 115 - 3.454 = 11.46$$

Mode: By inspection mode lies in the class 110–120

$$\text{Mode} = L + \frac{\Delta_1}{\Delta_1 + \Delta_2} \times i$$

$$L = 110,\ \Delta_1 = (50 - 42) = 8,$$

$$\Delta_2 = (50 - 45) = 5,\ i = 10$$

$$\therefore \quad \text{Mode} = 110 + \frac{8}{8+5} \times 10 = 110 + 6.15 = 116.15$$

$$\text{Standared Deviation } s = \sqrt{\frac{\Sigma fd^2}{N} - \left(\frac{\Sigma fd}{N}\right)^2} \times i \sqrt{\frac{773}{240} - \left(\frac{-85}{240}\right)^2} \times 10$$

$$= \sqrt{3.221 - 0.125} \times 10 = 1.7595 \times 10\ 17.595.$$

$$\text{Coeff. of Sk} = \frac{111.46 - 116.15}{17.595} = \frac{-4.69}{17.595} = -\ 0.266$$

Example 2:

Calculate coefficient of skewness based on quartiles and median from the following data:

Variable	*Frequency*	*Variable*	*Frequency*
0-10	*12*	*40-50*	*22*
10-20	*16*	*50-60*	*15*
20-30	*26*	*60-70*	*7*
30-40	*38*	*70-80*	*4*

Solution:

The Bowley's coefficient of skewness given by the following formula is based on quartiles and median:

$$Sk_B = \frac{Q_3 + Q_1 - 2\ \text{Med.}}{Q_3 - Q_1}$$

Calculation of Bowley's Coefficient of Skewness

Variable	f	c.f.
0-10	12	12
10-20	16	28
20-30	26	54
30-40	38	92
40-50	22	114
50-60	15	129
60-70	7	136
70-80	4	140

Q_1 = Size of $\frac{N}{4}$th item = $\frac{140}{4}$ = 35th item. It lies in the class 20-30

$$Q_1 = L + \frac{N/4 - c.f.}{f} \times i$$

L = 20, N/4 = 35, c.f. = 28, f = 26, i 10

$$Q_1 = 20 + \frac{35 - 28}{26} \times 10 = 20 + 2.69 = 22.69$$

Q_3 = Size of $\frac{3N}{4}$th item = $\frac{3 \times 140}{4}$ =105th item.

If lies in the class 40-50.

$$Q_3 = L + \frac{3N/4 - c.f.}{f} \times i$$

L = 40, 3N/4 = 105, c.f. = 92, f = 22, i = 10

$$Q_3 = 40 + \frac{105 \times - 92}{22} \times 10 = 40 + 5.91 = 45.91$$

Med. = Size of $\frac{N}{2}$th item = $\frac{140}{2}$ = 70th item. It lies in the class 30-40.

$$\text{Med.} = L + \frac{N/2 - c.f.}{f} \times i$$

L = 30, N/2 = 70, c.f. = 54, f = 38, i = 10

$$\text{Med.} = 30 + \frac{70 - 54}{38} \times 10 = 30 + 4.21 = 34.21$$

$$\text{Coeff. of Sk} = \frac{45.91 + 22.69 - 2\,(34.21)}{45.91 - 22.69} = \frac{68.6 - 68.42}{23.22} = 0.008.$$

Example 3:

You are given the position in a factory before and after the settlement of an industrial dispute. Comment on the gains or losses from the point of view of workers and that of management.

	Before	***After***
No. of workers	*2,400*	*2,350*
Mean wages (Rs.)	*45.5*	*47.5*
Median wages (Rs.)	*48.0*	*45.0*
Standard Deviation (Rs.)	*12.0*	*10.0*

Solution:

The following comments can be made on the basis of the information given.

(i) By comparing the total wage bill we can comment on the increase or decrease in the level of wages.

Total wage bill before the settlement of dispute = 2,400 × 46.5

= Rs. 1,09, 200

Total wage bill after the settlement of dispute = 2,350 × 47.5

= Rs. 1,11, 625.

Hence the total wage bill has gone up after the settlement of dispute even though the number of workers has decreased from 2,400 to 2,350. This means that the average wage is now higher. This is definitely a gain to the workers.

Conversely, we cannot say that increased wage bill is necessarily a loss to management because if it results in greater efficiency of workers and therefore higher productively, it would be a positive gain to management also.

(ii) Median before settlement of the dispute was 48 and after settlement it is 45. This means that formerly 50% of workers used to get wages above Rs. 48 and now they get only above Rs. 45.

(iii) By comparing the coefficient of variation before and after the settlement of dispute we can comment on the distribution of wages.

Coefficient of variation before the settlement of dispute

$$\text{C.V.} = \frac{\sigma}{X} \times 100, \text{ where } \sigma = 12,\ \overline{X} = 45.5$$

$$\therefore \quad \text{C.V.} = \frac{12}{45.5} \times 100 = 26.27$$

Coefficient of variation after the settlement of dispute $\sigma = 12$, $\overline{X} = 47.5$

$$\therefore \quad \text{C.V.} = \frac{10}{47.5} \times 100 = 21.05$$

Since the value of the coefficient of variation has decreased from 26.27 to 21.05 there is sufficient evidence to conclude that wages are more uniformly distributed after the settlement of dispute or, in other words, there is lesser inequality in the distribution of wages after the dispute is settled.

(iv) By comparing skewness we can comment upon the nature of the distribution.

Coefficient of skewness before the settlement of dispute

$$SK_P = \frac{3(\overline{X} - \text{Med.})}{\sigma} = \frac{3(45.5 - 48)}{12}$$

$$= \frac{-7.5}{12} = 0.625$$

Coefficient of skewness after the settlement of dispute

$$Sk_P = \frac{3(47.5 - 45)}{10} = \frac{7.5}{10} = +\ 0.75$$

Thus, the distribution is positively skewed after the settlement of dispute whereas it was negatively skewed before the settlement of dispute. This suggests that the number of workers getting low wages has increased considerable and that of workers getting high wages fallen, though the actual wage of workers has increased.

Example 4:

Calculate Bowley's coefficient of skewness for the data given below:

Weight (in lbs.)	*Number of Students*	*Weights (in lbs.)*	*Number of students*
Below 99	*1*	*150-159*	*65*
100-109	*14*	*160-169*	*34*
110-119	*66*	*170-179*	*12*
120-129	*122*	*180-189*	*5*
130-139	*145*	*190-199*	*2*
140-149	*121*	*200 and over*	*2*

Solution:

Calculation of Bowley's Coefficient of Skewness

Weight (in lbs.)	f	c.f.
Below 99	1	1
100-109	14	15
110-119	66	81
120-129	122	203
130-139	145	348
140-149	121	469
150-159	65	534
160-169	34	568
170-179	12	580
180-189	5	585
190-199	2	587
200 and over	2	589

$$\text{Coeff. of Sk (Bowley)} = \frac{Q_3 + Q_1 - 2\,\text{Med}}{Q_3 - Q_1}$$

$$\text{Med} = \text{Size of } \frac{N}{2}\text{th item} = \frac{589}{2} = 294.5\text{th item}$$

Median lies in the class 130–139. But the real limit of this class is 129.5–139.5

$$\text{Med.} = L + \frac{\frac{N}{2} - \text{c.f.}}{f} \times i$$

$L = 129.5$, $N/2 = 294.5$, c.f. $= 203$, $f = 145$, $i = 10$

$$\text{Med} = 129.5 + \frac{294.5 - 203}{145} \times 10 = 129.5 + 6.31 = 135.81$$

$$Q_1 = \text{Size of } \frac{N}{4}\text{th item} = \frac{589}{4} = 147.25\text{th item}$$

Q_1 lies in the class 120-129. But the real limit of this class is 119.5 to 129.5

$$Q_1 = L + \frac{\frac{N}{4} - \text{c.f.}}{f} \times i$$

$L = 119.5$, $N/4 = 147.25$, c.f. $= 81$, $f = 122$, $i = 10$

$$Q_1 = 119.5 + \frac{147.25 - 81}{122} \times 10 = 119.5 + 5.43 = 124.93$$

$$Q_3 = \text{Size of } \frac{3N}{4}\text{th item} = \frac{3 \times 589}{4} = 441.75\text{th item.}$$

Q_3 lies in the class 140-149

But the real limit of this class is 139.5-149.5

$$Q_3 = L + \frac{\frac{3N}{4} - c.f.}{f} \times i$$

$$L = 139.5,\ \frac{3N}{4} = 441.75,\ c.f. = 348,\ f = 121,\ i = 10$$

$$Q_3 = 139.5 + \frac{44.175 - 348}{121} \times 10 = 139.5 + 7.75 = 147.25$$

$$\text{Coeff. of Sk} = \frac{147.25 + 124.93 - 2(135.81)}{139.5 - 124.93}$$

$$= \frac{272.18 - 271.62}{14.57} = \frac{0.56}{14.57} = 0.038$$

Example 5:

Find Bowley's coefficient of skewness for the following frequency distribution:

No. of children per family	*0*	*1*	*2*	*3*	*4*	*5*	*6*
No. of families	*7*	*10*	*16*	*25*	*18*	*11*	*8*

Solution:

Calculation of Bowley's Coefficient of Skewness

No. of children per family X	**N. of families f**	**c.f.**
0	7	7
1	10	17
2	16	33
3	25	58
4	18	76
5	11	87
6	8	95

$$Sk_B = \frac{Q_3 + Q_1 - 2\text{ Med.}}{Q_3 - Q_1}$$

$$Q_1 = \text{Size of } \frac{N+1}{4}\text{th item} = \frac{95 \times 1}{4} = 24\text{th item, Hence } Q_1 = 2$$

$$Q_3 = \text{Size of } \frac{(N+1)}{4}\text{th item} = \frac{3 \times 95}{4} = 72\text{th item}$$

Size of 72th item is 4, Hence $Q_3 = 4$

$$\text{Med.} = \text{Size of } \frac{N+1}{2}\text{th item} = \frac{96}{2} = 48\text{th item.}$$

Size of 48th item is 3. Hence median = 3

$$Sk_B = \frac{4+2-(3)}{4-2} = \frac{0}{2} = 0.$$ **Ans.**

Example 6:

Calculate Karl Pearson's coefficient of skewness:

Variable	*Frequency*	*Variable*	*Frequency*
70-80	*11*	*30-40*	*21*
60-70	*22*	*20-30*	*11*
50-60	*30*	*10-20*	*6*
40-50	*35*	*0-10*	*5*

Solution:

Rearranging the given data in ascending order and calculating Karl Pearson's coefficient of skewness.

Calculation of Coefficient of Skewness

Class	**m.p.** m	**f**	**(m–35)/10** d	**fd**	**fd²**
0-10	5	11	–3	–33	99
10-20	15	22	–2	–44	88
20-30	25	30	–1	–30	30
30-40	35	35	0	0	0
40-50	45	21	+1	+21	21
50-60	55	11	+2	+22	44
60-70	65	6	+3	+18	54
70-80	75	5	+4	+20	80
		N = 141		$\Sigma fd = -26$	$\Sigma fd^2 = 416$

$$\text{Coeff. of Sk} = \frac{\text{Mean} - \text{Mode}}{\sigma}$$

Mean: $\overline{X} = A + \frac{\Sigma fd}{N} \times i$

$A = 35, \Sigma fd = -26, N = 141, i = 10$

$$\therefore \quad \overline{X} = 35 - \frac{26}{141} \times 10 = 35 - 1.844 = 33.156$$

Mode: By inspection mode lies in the class 30-40.

$$Mo = L + \frac{\Delta_1}{\Delta_1 + \Delta_2} \times i$$

$L = 30, \Delta_1 = (35 - 30) = 5,$

$\Delta_2 = (35 - 21) = 14, i = 10$

$$Mo = 30 + \frac{5}{5 + 14} \times 10 = 30 + 2.632 = 32.\ 632$$

$$S.D. = \sigma = \sqrt{\frac{\Sigma fd^2}{N} - \left(\frac{\Sigma fd}{N}\right)^2} \times i = \sqrt{\frac{416}{141} - \left(\frac{-26}{141}\right)^2} \times 10$$

$$= \sqrt{2.95 - .034} \times 10 = 1.708 \times 10 = 17.08$$

$$\text{Coeff. of Sk} = \frac{33.156 - 32.632}{17.08} = 0.031.$$

Example 7:

Calculate Pearson's coefficient of skewness:

x:	*12.5*	*17.5*	*22.5*	*27.5*	*32.5*	*37.5*	*42.5*	*47.5*
f:	*28*	*42*	*54*	*108*	*129*	*61*	*45*	*33*

Solution:

Calculation of Coefficient of Skewness

x	f	(x – 27.5)/5 d	fd	fd²
12.5	28	–3	–84	252
17.5	42	–2	–84	168
22.5	54	–1	–54	54
27.5	108	0	0	0
32.5	129	+1	+129	129
37.5	61	+2	+122	244
42.5	45	+3	+135	405
47.5	33	+4	+132	528
	N = 500		Σfd = 296	Σfd² = 1780

$$\text{Coeff. of Sk} = \frac{\text{Mean} - \text{Mode}}{\sigma}$$

Mean $$\overline{X} = A + \frac{\Sigma fd}{N} \times i$$

$$A = 27.5,\ \Sigma fd = 296,\ \ N = 500,\ i = 5.$$

$$\overline{X} = 27.5' + \frac{296}{500} \times 5 = 30.46$$

Mode: Since the maximum frequency is 129, the corresponding value of X, *i.e.*, 32.5 is the modal value.

S.D. $$\sigma = \sqrt{\frac{\Sigma fd^2}{N} - \left(\frac{\Sigma fd}{N}\right)^2} \times i$$

$$\Sigma fd^2 = 1780,\ \ N = 500,\ \Sigma fd = 296,\ \ i = 5$$

$$\sigma = \sqrt{\frac{1780}{500} - \left(\frac{296}{500}\right)^2} \times 5 = \sqrt{3.56 - .35} \times 5 = 8.96$$

$$\text{Coeff. of Sk.} = \frac{30.46 - 32.5}{8.96} = \frac{-2.04}{8.96} = 0.228$$ **Ans.**

MOMENTS

'Moment' is a familiar mechanical term which refers to the measure of a force with respect to its tendency to provide rotation. The strength of the tendency depends on the amount of force and the distance from the origin of the point at which the force is exerted. If a number of forces. F_1, F_2, F_n at distances X_1. X_2. X_n are applied. The moment of the first force about the origin is F_1X_1, the moment of the second fore is F_2X_2 etc. These moments are additive so that ΣFX is the total moment about the origin. If the total moment is divided by the total force, the quotient is termed "a moment". The formula is $\frac{\Sigma FX}{N}$ where $N = \Sigma F$ is the total force.

The following diagram shows the above fact.

In on both the sides of the fuierum the forces (weight × distance) are equal, then there will be a balance. At a balanced position the positive product equals the negative product.

However the term moment as used in physics has nothing to do with the moment used in statistics, the only analogy being that in statistics we talk of moment of random variable about some point. The moments in statistics are used to describe the various characteristics of a frequency distribution like central tendency, variation, skewness and kurtosis.

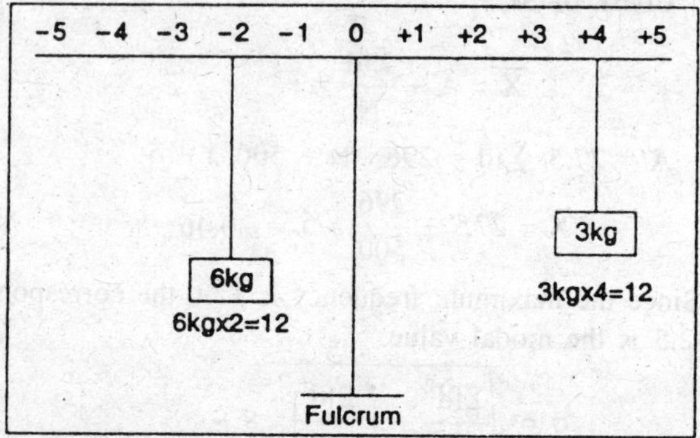

It can be seen that the formula for a moment coefficient is identical with that for an arithmetic mean. This identity has led statisticians to speak of the arithmetic mean as the "*first moment about the origin*". Technically the mean is moment coefficient and not a total moment, but in the case of frequency curves, with which Textbook of Sampling and Attributes is primarily concerned, the total frequency N is generally taken as unity, os the total moment and the moment coefficient are identical. In any case it has become customary in statistics to speak of the mean $\overline{X} = \frac{\Sigma FX}{N}$ as the first moment about the origin, and distinction between the total moment and moment coefficient is ignored. The concept of moments is also extended to higher powers. The statistical definition of the term 'moment' is given below:

Let the symbol x be used to represent the deviation of any item in a distribution from the aribution. If we take the mean of the first power of the deviations we get the first moment about the mean: the mean of the squares of the deviations gives us the second moment about the mean; the mean of the cubes of the deviations gives us the second moment about the mean; and so on. The moments about mean called the 'central moment' are denoted by Greek letter μ (read as mu); thus μ_1 stands for first moment about mean.μ_2 stands for second moment about mean. etc, Symbolically.

$$\mu_1 = \frac{\Sigma(X - \overline{X})}{N} \Rightarrow \frac{\Sigma x}{N}$$

(since sum of deviation of items from arithmetic mean is always zero. μ_1 would always be zero)

$$\mu_2 = \frac{\Sigma(X - \overline{X})^2}{N} \Rightarrow \frac{\Sigma x^2}{N}$$

$(\mu_2 = \sigma^2 = \Rightarrow \sigma = \sqrt{\mu_2})$

$$\mu_3 = \frac{\Sigma(X-\overline{X})^3}{N} \Rightarrow \frac{\Sigma x^3}{N}$$

For frequency distribution

$$\mu_1 = \frac{\Sigma f(X-\overline{X})}{N} \Rightarrow \frac{\Sigma fx}{N}; \mu_2 = \frac{\Sigma f(X-\overline{X})^2}{N} \Rightarrow \frac{\Sigma fx^2}{N}$$

$$\mu_3 = \frac{\Sigma f(X-\overline{X})^3}{N} \Rightarrow \frac{\Sigma fx^3}{N}; \mu_4 = \frac{\Sigma f(X-\overline{X})^4}{N} \Rightarrow \frac{\Sigma fx^4}{N}$$

Moments can be extended to highter powers in a similar way, but generally in practice the first four moments suffice. Furthermore, as pointed out by Yule and Kendall, "moment of higher order, through important in theory, are so extremely sensive to sampling fluctuations that value calculated for moderate number of observations are quite unreliable and hardly ever repay the labour of computation".

Two important constants of a distribution are calculated from μ_2, μ_3, and μ_4. They are:

(i) β_1 (read as beta one) $= \dfrac{\mu_3^2}{\mu_2^3}$

(ii) β_2 (read as beta two) $= \dfrac{\mu_4}{\mu_2^2}$

β_1 meansuires skewness and β_2 kurtosis.

In a symmetrical distribution all odd moments, *i.e.*, μ_1. μ_3 etc. would always be zero. The reason is that if the curve is symmetical there will be a deviation below the mean which exactly equals each deviations above the mean and therefore, positive deviations and negative deviation will exactly balance out and when added will cancel out *i.e.*, $\Sigma(X - \overline{X})$ would always be zero. Of course, if the deviations are raised to even powers their sign will always be positive and they will no longer cancel out. But the sum of the odd powers will all be equal to zero on account of the cancellations. Thus. *odd moments are always zero in symmetrical distribution.* However, this rule does not hold true in asymmetrical distribution.

Moments about Arbitrary Origin

Where the actual mean is in fractions it is difficult to calculate moments by peopling the above formulae. In such a case we can first compute moments about an arbitrary origin 'A' and then convert these moments into moments about the actual mean. Moments about arbitrary origin, also called 'raw moments', are denoted by the symbol μ' to distinguish them from the moments about mean which are denoted by μ. Thus μ_1' would stand for the first moment about an arbitrary 'A' μ_2' for the second moment about an arbitrary origin A, and so on. The calculations shall be done as follows:

$$\mu_1' = \frac{\Sigma(X - A)}{N} = (\overline{X} - A); \quad \mu_2' = \frac{\Sigma(X - A)^2}{N}$$

$$\mu_3' = \frac{\Sigma(X - A)^3}{N}; \quad \mu_4' = \frac{\Sigma(X - A)^4}{N}$$

For a frequency distribution

$$\mu_1' = \frac{\Sigma(X - A)}{N} \Rightarrow \frac{\Sigma fd^2}{N} \Rightarrow \frac{\Sigma fd^2}{N} \times i \quad \left[\text{where } d = \frac{(X - A}{N}\right]$$

$$\mu_2' = \frac{\Sigma f(X - A)^2}{N} \Rightarrow \frac{\Sigma fd^2}{N} \Rightarrow \frac{\Sigma fd^2}{N} \times i^2$$

In the same way the formulae for 3rd and 4th moments can be written.

Conversion of Moments about an Arbitrary Origin into Moments about Mean or Central Moments

For the sake of simplicity in calculations moments are first calculated about an arbitrary origin. If we want to obtain moments about mean we can do so with the following relationship:

$$\mu_1 = \mu_1' = \mu_1' = 0; \; \mu_3 = \mu_3' - 3\mu_1'\,\mu_2' + 2(\mu_1')^3$$

$$\mu_2 = \mu_2' - (\mu_1')^2; \; \mu_4 = \mu_4' - 4\mu_1'\mu_3' + 6(\mu_1')^2\,(\mu_2') - 3(\mu_1')^4$$

Relationship between the moments about mean and in terms of moments about any arbitrary point and conversely

$$\mu_r = \frac{1}{N}\,\Sigma(X_i + \overline{X})^r, \; \mu_i' = \frac{1}{N}\,\Sigma(X_i - a)^2$$

$$X_i\;\overline{X} = (X_i - a) - (\overline{X} - a)$$

$$\mu_r = \frac{1}{N}\,\Sigma(X_i - d)^r$$

where $X_i = x_i - a$ and $d = \overline{X} - a$

Using Binomial theorem

$$= \frac{1}{N}[\Sigma X_i^r - {}^rC_1 d\Sigma X_i^{r-1} + {}^rC_2 d^2 \Sigma X_i^{r-2} \ldots + {}^rC_{r-1}\,(-d)^{r-1}\,\Sigma X_i + (-d)^r]$$

$$= \mu_r^1 - {}^rC_1\,d\mu'_{r-1} + {}^rC_2 d^3 \mu'_{r-2} + \ldots + (-1)^{r-1}\,{}_r d^{r-1}\,\mu_1' + (-1)^r\,d^r$$

$$\mu_r = \mu_r' - {}^rC_1\mu'_{1\,r-1} + {}^rC_1\mu_1'^2\,\mu'_{r-2} + (-1)^{r-1}\,(r-1)\,\mu_1''^r$$

Putting r = 1, 2, 3, 4 we get

$$\mu_1 = \mu_1' - \mu_1' = 0$$

$$\mu_2 = \mu_2' - (\mu_1')^2$$

$$\mu_3 = \mu_3' - 3\mu_1'\mu_2' + 3(\mu_1')^3$$

$$\mu_4 = \mu_4' - 4\mu_1'\mu_3' + 6(\mu_1')^2\ \mu_2' - 3(\mu_1')^4$$

Conversely.

$$\mu_1' = \frac{1}{N}\Sigma(X_i - a)^2$$

$$= \frac{1}{N}\Sigma(X - \overline{X} + \overline{x} - a)^r$$

$$\frac{1}{N}\ \Sigma(x_1' - d)^r \text{ where } x_i' = x_i - \overline{x} \text{ and } d = \overline{x} - a$$

$$= \frac{1}{N}\ \Sigma[x_1'^r + {}^rC_1\ d\Sigma x_i'^{r-1} + {}^rC_2 d^2\Sigma x_i'^{r-2} + {}^rC_{r-1}a^{r-1}\ \Sigma x_1' + a^r]$$

$$= \mu_r{}^rC_1\mu_r - 1 + d^r\ C_2\mu_r - 2d^2 + ... + C_2\mu_2 d^{r-2} + d^r$$

In particular

$$\mu_2' = \mu_2 + d^2$$

$$\mu_3' = \mu_3' - 3du_2 + d^3$$

$$\mu_4' = \mu_1 + 4d\mu_3 + 6d^2\mu_2 + d^4$$

Moments about zero

The moments about zero are often denoted by υ_1. υ_2, υ_3 and are obtained as follows:

$$\upsilon_1 = \frac{\Sigma fX}{N} \qquad \upsilon_2 = \frac{\Sigma fX^2}{N}$$

$$\upsilon_3 = \frac{\Sigma fX^3}{N} \qquad \upsilon_4 = \frac{\Sigma fX^4}{N}$$

Also:

The first moment about zero or $\upsilon_1 = A + \mu_1'$ or the mean*

The second moment about zero or $\upsilon_2 = \mu_2 + (\upsilon_1)^2$.

The third moment about zero or $\upsilon_3 = \mu_3 + 3\upsilon_1\upsilon_2 - 3\upsilon_1{}^3$

The fourth moment about zero or $\upsilon_4 = \mu_4 + 4\upsilon_1\upsilon_3 - 6\upsilon_1{}^2\upsilon_2 + 3\upsilon_1{}^4$

Purpose of Moments. The concept of moment is of great significance in statistical work. With the help of moments we can measure the central tendency of a set of observations, their variability, their asymmetry and the height of the peak their curve would make. Because of the great convenience in obtaining measures of the various characteristics of a frequency distribution, the calculation of the first four moments about the mean may well be made the first step in the analysis of a frequency distribution.

The Following is the summary of how moments help in analysing a frequency distribution:

	Moment	*What it measures*
1.	First moment about origin	Mean
2.	Second moment about the mean	Variance
3.	Third moment about the mean	Skewness
4.	Fourth moment about the mean	Kurtosis

SHEPPARD'S CORRECTION FOR GROUPING ERRORS

While calculating moments it is assumed that all the values of a variable in a class interval are concentrated at the centre of that interval (*i.e.*, mid-point). However, in practice, it is not so–the assumption is an approximation to facilitate calculations and it introduces some error which is known as *grouping error.* But for distributions of symmetrical or moderately a skew type and class intervals not greater than about one-twentieth of the range, the approximation may be very close to one. In other cases we should apply Sheppard's corrections to eliminate grouping error.

The moments that we have computed, which have not been corrected by Sheppard's process, are called the *crude moments* to distringuish them from the adjusted moments which we get by applying Sheppard's corrections. These corrections are:

$$m_2(\text{corrected}) = \mu_2(\text{uncorrected}) - i^2/12$$

$$m_4(\text{corrected}) = \mu_4(\text{uncorrected}) - \frac{1}{2}\, i^2\mu^2(\text{uncorrected}) + \frac{7}{240} i^4$$

there i is the width of the class interval.

The first and third moments need no correction.

Conditions fro Applying Sheppard's Corrections

The following conditions should be satisfied for the application of Sheppard's corrections:

(a) The correction is not applicable to J- or U-shaped distributions or even to the skew form.

(b) The correction should not be made unless the frequency is at least 1,000 otherwise the moments will be more affected by sampling errors than by grcuping errors.

(c) The frequencies should taper off to zero in both directions, *i.e.*, the curve should approach the base line gradually and slowly at each end of the distribution.

(d) The observations should relate to a continuous variable.

However, as pointed out by A.E. Wauth, the corrections are small and the statistician is Foolish to other with them if the original figures are rough approximations. But where we have continuous data with the characteristics described above and where the original measurements are reasonably precise, may well apply Sheppard's corrections to eliminate the grouping error.

Example 1:

The first four moments of a distribution about x = 2 are 1, 2.5, 5.5 and 16. Calculate the four moments about $\overline{X}$ *and about zero.*

(M.Com. Delhi Univ.; M.Com., M.D. Univ.; B.Com. (H). Delhi. 1999)

Solution:

We are given $\mu_1' = 1$, $\mu_2' = 2.5$, $\mu_3' = 5.5$. and $\mu_4' = 16$. From these moments about arbitrary origin, we can find out moments about means with the help of the following relationship:

$$\mu_2 = \mu_2' - (\mu_1')^2$$

$$\mu_3 = \mu_3' - 3\mu_1'\mu_2' + 2(\mu_1')^3$$

$$\mu_4 = \mu_4' - 4\mu_1'\mu_3' + 6(\mu_1')^2\,\mu_2' - 3(\mu_1')^4$$

Substituting the value

$$\mu_2 = 2.5 - (1)^2 = 1.5$$

$$\mu_3 = 5.5 - 3(1)\,(2.5) + 2(1)^2 = 5.5 - 7.5 + 2 = 0$$

$$u_4 = 16 - 4(1)\,(5.5) + 6(2.5)\,(1)^2 - 3(1)^4 = 16 - 22 + 15 - 3 = 6.$$

Thus moments about mean are $\mu_1 = 0$, $\mu_2 = 1.5$, $\mu_2 = 0$, $\mu_4 = 6$

Moment about zero:

Let moment about zero be denoted by v_1, v_2, v_3 etc.

The first moment about zero, *i.e.*, $v_1 = A + \mu_1'$ or mean

The second moment about zero *i.e.*, $v_2 = \mu_2 + v_1^2$

The third moment about zero *i.e.*, $v_3 = \mu_2 + 3v_1v_2 - 2v_1^3$

The fourth moment about zero *i.e.*, $v_4 = \mu_1 + 4v_1v_3 - 6v_1^2v_2 + 3v_1^4$

In this question,

$v_1 = 2 + 1 = 3;$

$V_3 = 0 + (3 \times 3 \times 10.5) - 2(3)^3 = 94.5 - 54 = 40.5$

$v_2 = 1.5 + (3)^2 = 10.5;\ v_4 = 6 + 4(2)(40.5) - 6(3)^2(10.5) + 3(3)^4$

$= 6 + 586 - 567 + 243 = 168.$

Example 2:

Calculate first four moments about the mean and also the value of β_1 *and* β_2 *from the following data:*

Marks:	*0-10*	*10-20*	*20-30*	*30-40*	*40-50*	*50-60*	*60-70*
No. of students:	*8*	*12*	*20*	*30*	*15*	*10*	*5*

Solution:

Calculation of First Four Moments and β_1 and β_2

Marks	**m.p. m**	**No. of students f**	**(m–35)/10 d**	**fd**	**fd^2**	**fd^3**	**fd^4**
0-10	5	8	–3	–24	72	–216	648
10-20	15	12	–2	–24	48	–96	192
20-30	25	20	–1	–20	20	–20	20
30-40	35	30	0	0	0	0	0
40-50	45	15	+1	+15	15	+15	15
50-60	55	10	+2	+20	40	+80	160
60-70	65	5	+3	+15	45	+135	405
		N = 100		$\Sigma fd = 18$	$\Sigma fd^2 = 240$	$\Sigma fd^3 = -102$	$\Sigma fd^4 = 1440$

$$\mu_1' = \frac{\Sigma fd}{N} \times i = \frac{-18}{100} \times 10 = -1.8$$

$$\mu_2' = \frac{\Sigma fd^2}{N} \times i^2 = \frac{240}{100} \times 100 = 240$$

$$\mu_3' = \frac{\Sigma fd^3}{N} \times i^3 = \frac{-102}{120} \times 1000 = -1020$$

$$\mu_4' = \frac{\Sigma fd^4}{N} \times i^4 = \frac{1440}{100} \times 10000 = 144000$$

Now we can convert moments about arbitrary origin to moment about mean.

$$\mu_2 = \mu_2' - (\mu_1')^2 = 240 - (-1.8)^2 = 240 - 3.24 = 236.76$$

$\mu_3 = \mu_3' - 3\mu_1'\mu_2' + 2\mu_1'^3 = -1020 - 3(1.8)(240) + 2(-1.8)^3$

$= 1020 + 1296 - 11.664 = 264.336$

$\mu_4 = \mu_4' - 4\mu_1'\mu_3' + 6(\mu_1')^2 \mu_2' - 3\mu_1'^4$

$= 144000 - 4(-1.8)^2(-1020) + 6(-1.8)^2(240) - 3(-1.8)^4$

$= 144000 - 7354 + 4665.6 - 31.4928 = 141290.11$

$$\beta_1 = \frac{\mu_3^2}{\mu_2^3} = \frac{(264.336)^2}{(236.76)^3} = 0.005;\ \beta_2 = \frac{\mu_4}{\mu_2^2} = \frac{141290.11}{(236.76)^2} = 2.521.$$

Example 3:

From the following data calculate moments about

(i) assumed mean 25, (ii) actual mean, (iii) moments about zero.

Variable:	*0-10*	*10-20*	*20-30*	*30-40*
Frequency:	*1*	*3*	*4*	*2*

Solution:

Calculation of Moments About Arbitrary Origin

Variable	m	f	(m–25)/10	fd	fd²	fd³	fd⁴
0-10	5	1	–2	–2	4	–8	16
10-20	15	3	–1	–3	3	–3	3
20-30	25	4	0	0	0	0	0
30-40	35	2	+1	+2	2	–2	2
		N = 10		Σfd = – 3	Σfd² = 9	Σfd³ = – 9	Σfd⁴ = 21

$$\mu_1' = \frac{\Sigma fd}{N} \times i = \frac{-3}{10} \times 10 - 3$$

$$\mu_2' = \frac{\Sigma fd^2}{N} \times i^2 = \frac{9}{10} \times (10)^2 = \frac{9}{10} \times 100 = 90$$

$$\mu_3' = \frac{\Sigma fd^3}{N} \times i^3 = \frac{-9}{10} \times 1{,}000 = -900$$

$$\mu_4' = \frac{\Sigma fd^4}{N} \times i^4 = \frac{21}{10} \times 10{,}000 = 21{,}000$$

Moments about mean $\mu_1 = 0$

$\mu_2 = \mu_2' - (\mu_1')^2 = 90 - (-3)^2 = 81$

$\mu_3 = \mu_3' - 3\mu_1'\mu_2' + 2\mu_1'^3 = -900 - 3[(-3)(90)] + 3(-3)^3$

$= -900 + 810 - 54 = -144$

$\mu_4 = \mu_4' - 4\mu_1'\mu_3' + 6(\mu_1)^2\ \mu_2' - 3\mu_1'^4$

$= 21{,}000 - 4(-3)\ (-900) + 6\ (-3)^2\ 90 - 3\ (-3)^4$

$= 21{,}000 - 10{,}800 + 4{,}860 - 243 = 14{,}817.$

Moments about Zero $v_1 = A + \mu_1'$ or the mean $= 25 - 3 = 22$

$v_2 = \mu_2 + (v_1)^2 = 81 + (22)^2 = 565$

$v_3 = \mu_3 + 3v_1^2v_2 - 2v_1^3 = -144 + 3(22)^2\ (565) - 2(22)^3$

$= -144 + 37{,}290 - 21{,}296 = 15{,}850$

$v_4 = \mu_4 + 4v_1v_3 - 6v_1^2v_2 + 3v_1^4$

$= 14817 + 4(22)\ (15850) - 6(22)^2\ (565) + 3(22)^4$

$= 14817 + 1394800 - 1640760 + 702768 = 741625.$

Example 4:

Analyse the frequency distribution by the method of moments:

2	*3*	*4*	*5*	*6*
1	*3*	*7*	*3*	*1*

Solution:

Calculation of First Four Moments

X	f	$(X - \overline{X})$ $\overline{X} = 4$ x	fx	fx²	fx³	fx⁴
2	1	–2	–2	4	–8	16
3	3	–1	–3	3	–3	3
4	7	0	0	0	0	0
5	3	+1	+3	3	+3	3
6	1	+2	+2	4	+8	16
	N = 15		Σfx = 0	Σfx² = 14	Σfx³ = 0	Σfx⁴ = 38

$$\mu_1 = \frac{\Sigma(X - \overline{X})}{N} \Rightarrow \frac{\Sigma fx}{N} = \frac{0}{15} = 0; \quad \mu_3 = \frac{\Sigma fx^3}{N} = 0$$

$$\mu_2 = \frac{\Sigma fx^2}{N} = \frac{14}{15} = 0.933; \quad \mu_4 = \frac{\Sigma fx^4}{N} = \frac{38}{15} = 2.533$$

$$\sigma = \sqrt{\text{var iance}} \Rightarrow \sqrt{\mu_2}; \qquad \sqrt{\mu_3} = \sqrt{0.933} = 0.966$$

$$\beta_1 = \frac{\mu_3^2}{\mu_2^3} = \frac{0}{(0.933)} = 0$$

In a symmetrical distribution β_1 is zero. Hence this distribution is symmetrical,

$$\beta_2 = \frac{\mu_4}{\mu_2^2} = \frac{2.532}{(0.933)} = \frac{2.533}{0.87} = 2.91$$

Since the value of β_2 is less then three, the distribution is platykurtic.

MEASURE OF SKEWNESS BASED ON MOMENTS

A measure of skewness is obtained by making use of the second and third moments about the mean. When the method of moments is applied β_2 is used as relative measure of skewness, β_1 is defined as:

$$\beta_1 = \frac{\mu_3^2}{\mu_2^3}$$

In a symmetrical distribution b_1 shall be zero. The greater the value of β_1 the more skewed the distribution. However, the coefficient β_1 as a measure of skewness has a serious limitation. β_1 as a measure of skewness cannot tell us about the direction of skewness, *i.e.*, whether it is positive or negative, this is for the simple reason that μ_1 being the sum of the cubes of the deviations from the mean may be positive or negative but μ_3^2 is always positive. Also μ being the variance is always positive. Hence $\beta_1 = \mu_3^2 / \mu_2^3$ is always positive. This drawback is removed if we calculate Karl pearson's γ_1 (pronounced as Gamma one). γ_1 is defined as the square root of β_1, *i.e.*,

$$\gamma_1 = \sqrt{\beta_1} = \frac{\mu_3}{\mu_2^3} = \frac{\mu_3}{\sigma_3}$$

The sign of skewness would depend upon the value of μ_2. If μ_3 is positive we will have positive skewness and if μ_3 is negative, we will have negative skewness. It is advisable to use γ_1 as a measure of skewness.

Kurtosis in Greek means "*bulginess*". In statistics kurtosis refers to the degree of flatness or peakedness in the region about the mode of a frequency curve. The degree of kurtosis of a distribution is measured relative to the peakedness of normal curve. In other words, measures of kurtosis tell us the extent to which a distribution is more peaked or flat-topped than the normal curve. If a curve is more peaked than the normal curve, it is called '*leptokurtic*'. In such a case items are more closely bunched around the mode. On the other hand, if a curve is more flat-topped than the normal curve, it is called '*platykurtic*'. The normal curve itself is known as '*mesokurtic*'. The condition of peakedness or flat-toppedness itself is known as kurtosis of cxcess. The concept of kurtosis is rarely used in elementary statistical analysis.

The following diagram illustrates the shape of three different curves mentioned above:

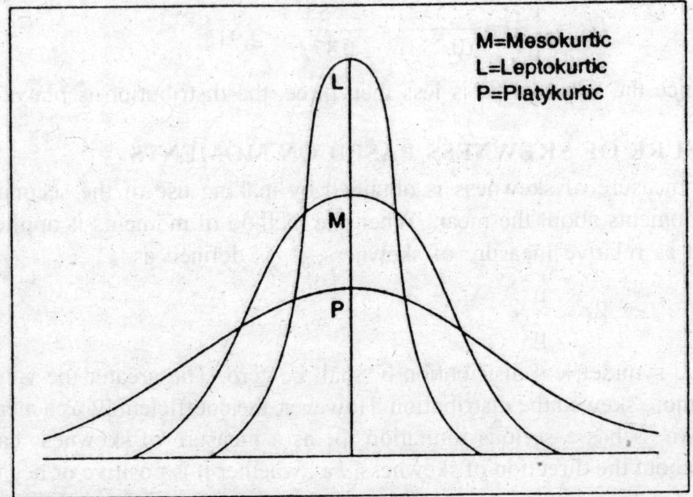

The above diagram clearly shows that these curves differ widely with regard to convexity an attribute which Karl Pearson referred to as 'kurtosis'. Curve M is a normal on and is called 'mesokurtic'. Curve L is more peaked than M and is called 'leptokurtic'. A leptokurtic curve has a narrower central portion and higher tails than does the normal curve. Curve P is less peaked (or more feat-topped) then curve M and is called 'platykurtic'. As may seen from the diagram, such a curve has a broader central and lower tails.

A famous British statistician William S. Gosset ("Student") has very humovosly pointed out the nature of these curves in the sentence, "Platykurtic curve, like the Platypus, are squat with short tails: leptokurtic curves are high with long tails like the kangaroos noted for leaping". Gosset's little sketched is reproduced below:

Measures of Kurtosis

The most important measure of kurtosis is the value of the coefficient β_2. It is defined as:

$\beta_2 = \frac{\mu_4}{\mu_2^2}$ where μ_4 = 4th moment and μ_2 = 2nd moment.

The greater the value of β_2, the more peaked the distribution.

For normal curve, the value of $\beta_2 = 3$. when the value of β_2 is greater then 3, the curve is more peaked than the normal curve. *i.e.*, leptokutic. When the value of β_3 less than 3 then the is curve is less peaked than the normal curve, *i.e.*, platykurtic. The normal curve and other curves with $\beta_2 = 3$ are called mesokurtic.

Sometimes γ_2, the derivative of β_2. is used as a measure of kurtiosis γ_2 is defined as

$$\gamma_2 = \beta_1 - 3.$$

For a normal distribution $\gamma_2 = 0$ is positive, the curve is leptokurtic and if γ_2 is negative, the curve is platykurtic.

Example 1:

From the information given below calculate Kari Pearson's coefficient of skewness and also quartile coefficient of skewness:

Measure	*Place A*	*Place B*
Mean	*150*	*140*
Median	*142*	*155*
S.D.	*30*	*55*
Third Quartile	*195*	*260*
First Quartile	*62*	*80*

Solution:

Calculation of Kari Pearson's coefficient of skewness

$$Skp = \frac{\overline{X} - \text{Mode}}{\sigma} \quad \overline{X} = 150, \sigma = 55$$

The value of mode is not given but can be ascertained from the relationship:

Mode : 3 Median – 2 Mean = (3 × 142) – (2 × 150) = 426 – 300 = 126

$$Skp = \frac{150-126}{30} = +0.8; \quad \textit{Place B}: \overline{X} = 140, \sigma = 55$$

Mode = 3 Median – 2 Mean = (3 × 55) – (2 × 140)

$$= 465 - 280 = 185; \quad Skp = \frac{140-185}{55} = -0.82$$

Quartile Coefficient of Skewness:

$$\textit{Place A}: Sk_B = \frac{Q_3 + Q_1 - 2\text{ Med.}}{Q_3 - Q_1}; \quad Q_3 = 195, Q_1 = 62, \text{Med} = 142$$

$$Sk_B = \frac{195 + 62 - (2 \times 142)}{195 - 62} = \frac{257 - 284}{133} = -0.203$$

$$\textit{Place B}: Sk_B = \frac{Q_3 + Q_1 - 2\text{ Med.}}{Q_3 - Q_1}; \quad Q_3 = 260, Q_1 = 80, \text{Med.} = 155$$

$$Sk_B = \frac{260 + 80 - (2 \times 155)}{260 - 80} = \frac{30}{180} = 0.167$$

Example 2:

The first four moments of a distribution about x = 4 are 1, 4, 10 and 45. Obtain the various characteristics of the distribution on the basis of the information given. Comment upon the nature of the distribution.

Solution:

We are given $\mu_1' = 1, \mu_2' = 4, \mu_3' = 10, \mu_4' = 45$ and A = 0. On the basis of this information we can calculate moment about mean and also determine the values of arithmetic mean, standard deviation, skewness and kurtosis.

Moment about mean : $\mu_1 = 0$; $\mu_2 = \mu_2' - (\mu_1')^2 = 4 - (1)^2 = 3$; $\mu_3 = \mu_3' - 3\mu_1'\mu_2' + 2(\mu_1)')^3 = 10 - 3(1)(4) + 2(1)^3 = 10 - 12+3 = 0$; $\mu_4 = \mu_4' - 4\mu_1'\mu_3' + 6\mu_2'(\mu_1')^2 - 3(\mu_1')^4 = 45 - 4(1)(10) + 6(4)(1)^2 - 3(1)^4 = 45 - 40 + 34 - 3 = 26$

Mean of the distribution $\overline{X} = V + \mu_1' = 4 + 1 = 5$

Standard Deviation : $\sigma = \sqrt{\mu_2} = \sqrt{3} = 1.732$

Skewness : $\beta_1 = \dfrac{\mu_3^2}{\mu_2^3} = \dfrac{(0)^2}{(3)^2} = 0$

Hence the distribution is symmetrical

Kurtosis : $\beta_2 = \frac{\mu_4}{\mu_2^2} = \frac{26}{(3)^2} = \frac{26}{9} = 2.89$

Since β_2 is less than 3, the distribution is platykurtic.

Example 3:

From the data given below calculate Karl Pearson's coefficient of skewness.

Age	*No. of persons*	*Age*	*No. of persons*
20-25	*50*	*40-45*	*150*
25-30	*70*	*45-50*	*120*
30-35	*80*	*50-55*	*70*
35-40	*180*	*55-60*	*50*

Solution:

Calculation of Coefficient of Skewness

Age	**m.p. m**	**No. of persons f**	**(m - 37.5)/5 d**	**fd**	**fd²**
20-25	22.5	50	–3	–150	450
25-30	27.5	70	–2	–140	280
30-35	32.5	80	–1	–80	80
35-40	37.5	180	0	0	0
40-45	42.5	150	+1	+150	150
45-50	47.5	120	+2	+240	480
50-55	52.5	70	+3	+210	630
55-60	57.5	50	+4	+200	800
		N = 770		Σfd = 430	Σfd² = 2870

$$Skp = \frac{\text{Mean} - \text{Mode}}{\sigma}$$

Mean : $\overline{X} = A = \frac{\sum fd}{N} \times i = 37.5 + \frac{430}{770} \times 5 = 37.5 + 2.79 = 40.29$

Mode : By inspection, it is not clear as to in which class mode lies. Hence we prepare an analysis table and grouping table.

Analysis Table

Age	Col.I	Col.II	Col.III	Col.IV	Col.V	Col.VI
20-25	50	120				
25-30	70		150	200		
30-35	80	260			330	
35-40	180		330			410
40-45	150	270		450		
45-50	120		190		340	
50-55	70	120				240
55-60	50					

Analysis Table

Class in which mode is expected to lie

Col. No.	35-40	40-45	45-50
I	1	1	1
II		1	
III	1	1	1
IV	1	1	1
V		1	
VI	1		
	4	5	3

It is clear from the analysis table that mode lies in the class 40-45.

$$\text{Mode} = L + \frac{\Delta_1}{\Delta_1 + \Delta_2} \times i$$

$L = 40,\ \Delta_1 = |\ 150 - 180\ | = 30,\ \Delta_2 = |\ 150 - 120\ | = 30,\ i = 5$

$$\therefore \quad \text{Mode} = 40 + \frac{30}{30 + 30} \times 5 = 40 + 2.5 = 42.5$$

$$\text{S.D.} : \sigma = \sqrt{\frac{\sum fd^2}{N} - \left(\frac{\sum fd}{N}\right)^2}\ \} \times i = \sqrt{\frac{2870}{770} - \left(\frac{430}{770}\right)^2} \times 5$$

$$= \sqrt{3.727 - 0.312} \times 5 = \sqrt{3.415} \times 5 = 1 = 1.848 \times 5 = 9.24$$

$\overline{X} = 40.29,\ \text{Mode} = 42.5,\ \sigma = 9.24$

$$\text{Skp} = \frac{40.29 - 42.5}{9.24} = \frac{-2.27}{9.24} = -0.239$$

Example 4:

Calculate the coefficient of skewness from the following data :

Mid-point :	*15*	*20*	*25*	*30*	*35*	*40*
Frequency :	*12*	*18*	*25*	*24*	*20*	*21*

Solution:

Calculation of Coefficient of Skewness

m.p. m	f	(m – 25)/5 d	fd	fd²
15	12	– 2	– 24	48
20	18	– 1	– 18	18
25	25	0	0	0
30	24	+ 1	+ 24	24
35	20	+ 2	+ 40	80
40	21	+ 3	+ 63	189
	N = 120		Σ fd = 85	Σfd^2 = 359

$$\text{Coeff. of Sk} = \frac{\text{Mean} - \text{Mode}}{\sigma}$$

$$\overline{X} = A + \frac{\Sigma fd}{N} \times i = 25 + \frac{85}{120} \times 5 = 25 + 3.542 = 28.542$$

$$\sigma = \sqrt{\frac{\Sigma fd^2}{N} - \left(\frac{\Sigma fd}{N}\right)^2} \times i \sqrt{\frac{359}{120} - \left(\frac{85}{120}\right)^2} \times 5$$

$$= \sqrt{2.992 - 5.02} \times 5 = 1.578 \times 5 = 7.89$$

Mode : The highest frequency is 25, mode lies corresponding to mid-point 25, *i.e.*, in the class 22.5 – 27.5.

$$Mo = L + \frac{\Delta_1}{\Delta_1 + \Delta_2} \times i; \quad L = 22.5,\ \Delta_1 = 25-18 = 7,\ \Delta_2 = 25-24 = 1,\ i = 5$$

$$Mo = 22.5 + 7/7 + 1 \times 5 = 22.5 + 4.375 = 26.875$$

$$Skp = \frac{28.542 - 26.875}{7.89} = \frac{1.667}{7.89} = 0.211.$$

There is a low agree of positive skewness.

Example 5:

The first four central moments of distribution are 0, 2.5, 0.7 and 18.75. comment on the skewness and kurtosis of the distribution.

Solution:

Testing Skewness

We are given $\mu_1 = 0$, $\mu_2 = 2.5$,

$$\mu_3 = 0.7 \text{ and } \mu_4 = 18.75$$

Skewness is measured by the coefficient β_1,

$$\beta_1 = \frac{\mu_3^2}{\mu_2^3},\ \mu_2 = 2.5\ \mu_3 = 0.7, \quad \therefore\ \beta_1 = \frac{(0.7)^2}{(2.5)^2} = 0.031$$

Since $\beta_1 = 0.031$, the distribution is slightly skewed, *i.e.*, it is not perfectly symmetrical.

Testing Kurtosis:

For testing kurtoris we compute the value of β_2. When a distribution is normal or symmetrical, $\beta_3 = 3$. When distribution is more peaked than the normal, β_2 is more than 3 and when it is less peaked than the normal, β_2 is less than 3.

$$\beta_2 = \frac{\mu_4}{\mu_2^2},\ \text{where } \mu_4 = 18.75\ \mu_2 = 2.5$$

$$\beta_2 = \frac{18.75}{(2.5)^2} = \frac{18.75}{6.25} = 3$$

Since β_2 is exactly three, the distribution is mesokurtic.

Example 6:

In a certain distribution the following results were obtained.

$\overline{X}$ = *45,* *Median* = *48* *Coeff. of Sk* = *– 0.4.*

The person who gave you the data failed to give the value of standared eviation and your are required to estimate it with the help of the available information.

Solution:

Since we are given the value of $\overline{X}$, Med. and Coefficient of skewness, we can find out standard deviation by putting these values in the formula for calculating skewness, *i.e.*,

$$\text{Coeff. of Sk} = \frac{3(\text{Mean} - \text{Median})}{\sigma}$$

$$-0.4 = \frac{3(45-48)}{\sigma} \quad -0.4\sigma = -9$$

$$\Rightarrow \qquad \sigma = \frac{9}{0.4} = 22.5$$

Hence the value of standard deviation is 22.5.

Example 7:

Compute the coefficient of skewness and kurtosis based on the following data:

X:	*4.5*	*14.5*	*24.5*	*34.5*	*44.5*	*54.5*	*64.5*	*74.5*	*84.5*	*94.5*
f:	*1*	*5*	*12*	*22*	*17*	*9*	*4*	*3*	*1*	*1*

Solution:

Calculation of Skewness and Kurtosis

X	f	(X–44.5)/10 d	fd	fd^2	fd^3	fd^4
4.5	1	–4	–4	16	–64	256
14.5	5	–3	–15	45	–135	406
24.5	12	–2	–24	48	–96	192
34.5	22	–1	–22	22	–22	22
44.5	17	0	0	0	0	0
54.5	9	+1	+9	9	+9	9
64.5	4	+2	+8	16	+32	64
74.5	3	+3	+9	27	+81	243
84.5	1	+4	+4	16	+64	256
94.5	1	+5	+5	25	+125	625
	N = 75		$\Sigma fd = -30$	$\Sigma fd^2 = 224$	$\Sigma fd^3 = -6$	$\Sigma fd^4 = 2,072$

$$\mu_1'^* = \frac{\Sigma fd}{N} = \frac{-30}{75} = -0.4; \qquad \mu_3'^* = \frac{\Sigma fd^2}{N} = \frac{-6}{75} = -0.08$$

$$\mu_2'^* = \frac{\Sigma fd^2}{N} = \frac{224}{75} = 2.99; \qquad \mu_4'^* = \frac{\Sigma fd^4}{N} = \frac{2,072}{75} = 27.63$$

$$\mu_2^* = \mu_2' (\mu_1')^2 = 2.99 - (0.4)^2 = 2.99 - 0.16 = 2.83$$

$$\mu_3^* = m_3' - 2\mu_1'\mu_3' + 2(\mu_1')^3 = -0.08 - 3(-0.4)(2.99) + 2(-0.4)^3$$

$$= -0.08 + 3.588 - 0.128 = 3.38$$

$$\mu_4^* = \mu_4' - 4\mu_1'\mu_3' + 6\mu_1'^2\mu_2' = 3(\mu_1')^4$$

$$= 27.63 - 4(-0.4)(-0.08) + 6(-0.4)^2(2.99) + 3(-0.4)^4$$

$$= 27.63 - 0.128 + 2.87 - 0.077 = 30.295$$

Skewness: $$\beta_1^* = \frac{\mu_3^2}{\mu_2^3} = \frac{(0.38)^2}{(2.83)^3} = \frac{11,424}{22.625} = 0.504.$$

For Kurtosis we have to compute the value of β_2

$$\beta_2{}^* = \frac{\mu_4}{\mu_2^2} = \frac{30.295}{(2.83)^3} = \frac{30.295}{8.01} = 3.782.$$

Since the value of b_2 is greater than 3, the then curve is more peach then the normal serve

Example 8:

In a frequency distribution the coefficient of skewness based on quartile is 0.6. If the sum of the upper and the lower quartiles is 100 and the median is 38, find the value of the upper quartile.

Solution:

We are given Coeff. of Sk = 0.6, $Q_1 + Q_3 = 100$, Med. = 38. We have to find Q_3.

$$Sk_B = \frac{Q_3 + Q_1 - 2\,\text{Med.}}{Q_3 - Q_1},\ 0.6 = \frac{100 - 2(38)}{Q_3 - Q_1}$$

Since $Q_3 + Q_1 = 100;$

$$Q_1 = 100 - Q_3$$

Putting value of Q_1 we get:

$$0.6 = \frac{100 - 76}{Q_3 - (100 - Q_3)} = \frac{24}{Q_3 - (100 - Q_3)}$$

$$0.6(2Q_3 - 100) = 24$$

$$1.2Q_3 - 60 = 24$$

$$\Rightarrow \quad 1.2Q_3 = 84$$

$$\Rightarrow \quad Q_3 = \frac{84}{1.2} = 70$$

Hence the value of upper quartile is 70.

Example 9:

From the data given below calculate the coefficient of variation: Karl Pearson's coefficient of skewness = 0.42

Arithmetic Mean = 86

Median = 80

Solution:

$$C.V. = \frac{\sigma}{X} \times 100$$

Since we are not given standard deviation, first we calculate its value by the use of skewness formula:

$$\text{Coeff. of Sk} = \frac{\overline{X} - \text{Mode}}{\sigma}$$

$$\text{Mode} = 3\text{Med.} - 2\overline{X} = 3(80) - 2(86) = 240 - 172 = 68$$

Coeff. of Sk = 42, $\overline{X}$ = 86, Mo = 68

Substituting the values $.42 = \frac{86 - 68}{\sigma}$

$$42\sigma = 18$$

$$\Rightarrow \quad \sigma = 42.86$$

$$\text{C.V} = \frac{42.86}{86} \times 100$$

$$= 49.84 \text{ percent.}$$

Example 10:

For a group of 20 items, $\Sigma X = 1452$, $\Sigma X^2 = 144280$ and mode = 63.7. Find the Pearsonian coefficient of skewness.

Solution:

$$Sk_P = \frac{\text{Mean} - \text{Mode}}{\sigma}$$

$$\text{Mean} = \frac{\Sigma X}{N} = \frac{1452}{20} = 72.6$$

$$\sigma = \sqrt{\frac{\Sigma X^2}{N} - (\overline{X})^2} = \sqrt{\frac{144280}{20} - (72.6)^2}$$

$$= \sqrt{7214 - 5270.76} = 44.082$$

$$Sk_P = \frac{72.6 - 63.7}{44.082} = 0.202.$$

Example 11:

Find the coefficient of skewness from the following information:

Difference of two quartiles = 8

Mode = 11

Sum of two quartiles = 22

Mean = 8

Solution:

$$\text{Coeff. of Sk} = \frac{Q_3 + Q_1 - 2\text{ Med.}}{Q_3 - Q_1}$$

From the given information we can find out median

$$\text{Mode} = 3\text{ med.} - 2\text{ mean}$$

$$3\text{ med.} - 2(8) = 11$$

$$3\text{ med} = 11 + 16 \text{ or med} = 9$$

$$Q_3 + Q_1 = 22,\ Q_3 - Q_1 = 8,\ \text{Med} = 9$$

Hence $\qquad \text{Coeff of Sk} = \dfrac{22 - 2(9)}{8} = 0.5.$

Example 12:

Using moments, calculate a measure of relative skewness and a measure of relative kurtosis for the following distribution and comment on the result obtained:

Daily wages (in rupees)	***No. of workers***
70 but below 90	*8*
90 but below 110	*11*
110 but below 130	*18*
130 but below 150	*9*
150 but below 170	*4*

Solution:

To find first four Central Moments and β_1 and β_2

Daily wages (Rs.)	**m.p. m**	**f**	**(m–120)/10 d**	**fd**	**fd²**	**fd³**	**fd⁴**
70-90	80	8	–2	–16	32	–64	128
90-110	100	11	–1	–11	11	–11	11
110-130	120	18	0	0	0	0	0
130-150	140	9	+1	+9	9	+9	9
150-170	160	4	+2	+8	16	+32	64
		N = 50		$\Sigma fd = -10$	$\Sigma fd^2 = 68$	$\Sigma fd^3 = -34$	$\Sigma fd^4 = 212$

$$\mu_1' = \frac{\Sigma fd}{N} \times i = \frac{-10}{50} \times 10 = -2,$$

$$\mu_2' = \frac{\Sigma fd^2}{N} \times i^2 = \frac{68}{50} \times 100 = 136$$

$$\mu_3' = \frac{\Sigma fd^3}{N} \times i^3 = \frac{-34}{50} \times 1000 = -680,$$

$$\mu_4' = \frac{\Sigma fd^4}{N} \times i^4 = \frac{212}{50} \times 10000 = 42400$$

Convert moments about arbitrary origin to moments about mean

$$\mu_2 = \mu_2' - (\mu_1')^2 = 136 - (-2)^2 = 132$$

$$\mu_3 = \mu_3' - 3\mu_1'\mu_2' + 2\mu_1'^3 = -680 - 3(-2)(136) + 2(-2)^3$$

$$= -680 + 816 - 16 = 120$$

$$\mu_4 = \mu_4' 4\mu_1'\mu_3' + 6(\mu_1')^2 \mu_2' - 3\mu_1'^4$$

$$= -42400 - 4(-2)(-680) + 6(-2)^2 (136) - 3(-2)^4$$

$$42400 - 5440 + 3264 - 48 = 40176$$

$$\beta_1 = \frac{\mu_3^2}{\mu_2^3} = \frac{(120)^2}{(132)^3} = \frac{14400}{2299968} = 0.006$$

$$\beta_2 = \frac{\mu_4}{\mu_2^2} = \frac{40176}{(132)^2} = \frac{40176}{17424} = 2.306$$

Example 13:

The daily wages earned by one hundred workers of a factory are set out in the following table:

Weekly wages (Rs.)	***No. of workers***	***Weekly wages (Rs.)***	***No. of workers***
12.5-17.5	*12*	*37.5-42.5*	*10*
17.5-22.5	*16*	*42.5-47.5*	*6*
22.5-27.5	*25*	*47.5-52.5*	*3*
27.5-32.5	*14*	*52.5-57.5*	*1*
32.5-37.5	*13*		

(i) Calculate the three quartiles of the above distribution.

(ii) Find the absolute measures of dispersion and skewness based on these quartiles.

(iii) Interpret all the five measures, that you have calculated in (i) and (ii) above.

Solution:

Calculations of Quartiles and Coefficient of Skewness

Daily wages (Rs.)	f	c.f.	Weekly wages (Rs.)	f	c.f.
12.5-17.5	12	12	37.5-42.5	10	90
17.5-22.5	16	28	42.5-47.5	6	96
22.5-27.5	25	53	47.5-52.5	3	99
27.5-32.5	14	67	52.5-57.5	1	100
32.5-37.5	13	80			

(i) Q_1 = Size of $\frac{N}{4}$th item = $\frac{100}{4}$ = 25th item

Q_1 lies in the class 17.5 – 22.5, $Q_1 = L + \frac{N/4 - c.f.}{f} \times i$

$L = 17.5$, $N/4 = 25$, c.f. = 12, f = 16, i = 5

$$Q_1 = 17.5 + \frac{25 - 12}{16} \times 5 = 17.5 + 4.06 = 21.56$$

Q_2 = Size of $\frac{2N}{4}$th item = $\frac{2 \times 100}{4}$ = 50th item

Q_2 lies in the class 22. 5–27.5, $Q_2 = L + \frac{2N/4 - c.f.}{f} \times i$

$L = 22.5$, $2N/4$, = 50,

c.f. = 28, f = 25, i = 5

$$Q_2 = 22.5 + \frac{50 - 28}{25} \times 5 = 22.5 + 4.4 = 26.9$$

Q_3 = Size of $\frac{3N}{4}$th item = $\frac{3 \times 100}{4}$ = 75th item

Q_3 lies in the class 32.5–37.5:

$$Q_3 = L + \frac{3N/4 - c.f.}{f} \times i$$

$L = 32.5$, $3N/4 = 75$,

c.f. = 67, f = 13, i = 5

$$Q_3 = 32.5 + \frac{75 - 67}{13} \times 5 = 32.5 + 3.08 = 35.58$$

(ii) Absolute measure of dispersion and skewness

Dispersion: $Q.D = \frac{Q_3 - Q_1}{2} = \frac{35.58 - 21.56}{2} = 7.01$

Skewness: Absolute skewness

$= Q_3 + Q_1 - 2\text{ Med.} = 35.58 + 21.56 - (2 \times 26.9) = 57.14 - 53.8 = 3.34$

(iii) *Significance of the Measures*

Quartile of the rank of $\frac{N}{4}$, the value where of Rs. 21.56 signifies that 25% of the workers in the factory received wages that were less than Rs. 21.56.

Quartile of the rank of $\frac{N}{2}$, the value where of is Rs. 26.9, signifies that 50% of the workers received wages that were less than Rs. 26.9 and 50% received wages that were more than this amount.

Quartile of the rank of $\frac{3N}{4}$, the value where of is Rs. 35.58, signifies that 25% of workers received more than this amount.

The dispersion calculated on this basis of the quartile–Quartile deviation in relation to median–will disclose whether of not the distribution is 'normal'. Thus, if median + Q.D. is equal to Q_3, and median – Q.D. is equal to Q_1, the distribution would be regarded as normal or symmetrical.

In this question. Median + Q.D. = (26.9 + 7.01) = 33.91, and Median Q.D. = (26.9 – 7.01) = 19.89. Since the figures are very different from Q_2 and Q_1, it can be inferred that the distribution is not 'normal'.

The absolute measure of skewness, the value where of is Rs. 3.34, indicates that there is a large concentration of items towards Q_3, *i.e.*, the distribution is positively skewed. Such an asymmetrical distribution, when graphically represented, would be to tail off towards the right.

Example 14:

For a moderately skewed data, the arithmetic mean is 200, the coefficient of variation is 8 and Karl Pearson's coefficient of skewness is 0.3. Find the mode and the median.

Solution:

We are given that

$$\overline{X} = 200,\ C.V. = 8,\ Sk_P = 0.3,$$

We have to find mode and median.

From the formula of coefficient of variation, we can determine the standard deviation

$$\text{C.V.} = \frac{\sigma}{X} \times 100 \qquad 8 = \frac{\sigma}{200} \times 100$$

$$\Rightarrow \qquad \sigma = 16$$

Now from the formula of coefficient of skewness, we can determine mode

$$\text{Coeff. of Sk} = \frac{\text{Mean} - \text{Mode}}{\sigma}$$

$$0.3 = \frac{200 - \text{Mode}}{46}$$

$$4.8 = 200 - \text{Mode or Mode} = 195.2.$$

The value of median can now be determined from the formula:

$$\text{Mode} = 3\ \text{Med.} - 2\overline{X}$$

$$195.2 = 3\ \text{Med.} - 2\ (200)$$

$$3\ \text{Med.} - 400 = 195.2$$

$$3\ \text{Med} = 595.2 \text{ or Median} = 198.4$$

Hence the mode and median are 195.2 and 198.4 respectively.

Example 15:

The following table gives the distribution of income of households based on hypothetical data:

Income (Rs.)	***% of households***	***Income (Rs.)***	***% of households***
Under 100	*7.2*	*500-599*	*14.9*
100-199	*11.7*	*600-699*	*10.4*
200-299	*12.1*	*700-799*	*9.0*
300-399	*14.8*	*1000 and above*	*4.0*
400-499	*15.9*		

(i) What are the problems involved in computing standard deviation from the above data?

(ii) Compute a suitable measure of dispersion.

(iii) would the skewness of above distribution be affected if the income of everyone was increased by a certain proportion?

Solution:

(i) Since it is an open-end distribution, for calculating standard deviation we shall have to make an assumption about the lower limit of the first class and the upper limit of the last class.

(ii) Suitable measure of dispersion in this case shall be the quartile deviation,

Calculation of $Q.D. = \frac{Q_3 - Q_1}{2}$

Income (Rs.)	% of household f	c.f.
Under 100	7.2	7.2
100-199	11.7	18.9
200-299	12.1	31.0
300-399	14.8	45.8
400-499	15.9	61.7
500-599	14.9	76.6
600-699	10.4	87.0
700-799	9.0	96.0
1000 and above	4.0	100.0

Q_1 = Size of $\frac{N}{4}$th item $= \frac{100}{4}$ = 25th item, Q_1 lies in the class 200–299. But the real limit of this class is 199.5–299.5

$$Q_1 = L + \frac{N/4 - c.f.}{f} \times i = 199.5 + \frac{25 - 18.9}{12.1} \times 100$$

$$= 199.5 + 50.41 = 249.91$$

Q_3 = Size of $\frac{3N}{4}$th item $= \frac{300}{4}$ = 75th item. Q_3 lies in the class 500–599. But the real limit of this class is 499.5–599.5.

$$Q_3 = L + \frac{3N/4 - c.f.}{f} \times i = 499.5 + \frac{75 - 61.7}{14.9} \times 100$$

$$= 499.5 + 80.26 = 588.76.$$

Since Q_1 = 249.91 and Q_3 = 588.76 quartile deviation shall be

$$Q.D. = \frac{588.76 - 249.91}{2} = 169.42.$$

(iii) If the income of every one is increased by a certain proportion, the skewness would not be affected.

Example 16:

You are given the weekly wages of 30 workers working in a household factory.

Wages (Rs.)			
350.4	*375.8*	*380.2*	*420.7*
469.2	*392.4*	*420.0*	*460.2*
578.3	*492.4*	*536.4*	*625.2*
633.2	*475.3*	*375.8*	*608.2*
592.6	*576.5*	*517.2*	*594.5*
462.7	*415.8*	*479.2*	*588.6*
572.5	*517.9*	*537.6*	*493.4*
708.5	*727.9*		

Form a frequency distribution of the above data taking the first class as 300-400 and from the frequency distribution so formed, calculate Karl Pearson's coefficient of skewness.

Solution:

Formation of frequency distribution

Wages (Rs.)	Tally Bars	Frequency
300-400	III	5
400-500	III III	10
500-600	III III	10
600-700	III	3
700-800	II	2

Since it is a bimodal ser

ies, the following formula for calculating skewness shall be applied :

$$\text{Skp} = \frac{3(\text{Mean} - \text{Median})}{\sigma}$$

Calculation of Coefficient of Skewness

Wages (Rs.)	m.p. m	f	(m-550)/100 d	fd	fd^2	c.f.
300-400	350	5	–2	–10	20	5
400-500	450	10	–1	–10	10	15
500-600	550	10	0	0	0	25
600-700	650	3	+1	+3	3	28
700-800		2	+2	+4	8	30
		N = 30		Σfd = –13	Σfd^2=41	

$$\overline{X} = A + \frac{\sum fd}{N} \times i = 550 \frac{13}{30} \times 100 = 550 - 43.33 = 506.67$$

Median : Med = Size of $\frac{N}{2}$ in item = $\frac{30}{2}$ = 15th item.

It lies in the class 400-500.

$$\text{Med.} = L + \frac{N/2 - c.f.}{f} \times i$$

$$L = 400,\ N/2 = 15,$$

$$c.f. = 5,\ f = 10,\ i = 100$$

$$\text{Med.} = 400 + \frac{15-5}{10} \times 100 = 400 + 100 = 500$$

$$\sigma = \sqrt{\frac{\sum fd^2}{N} - \left(\frac{\sum fd}{N}\right)^2} \times i\ \sigma = \sqrt{\frac{41}{30} - \left(\frac{-13}{30}\right)^2} \times 100$$

$$= \sqrt{1.367 - .188} \times = \sqrt{1.179} \times 100 = 1.086 \times 100 = 108.6$$

$\overline{X}$ = 506.67, Median = 500, σ 108.6

$$\text{Coeff. of Sk} = \frac{3(506.67 - 500)}{108.6} = \frac{20.01}{108.6} = +\ 0.184\ .$$

MISCELLANEOUS EXAMPLES

Example 1:

A frequency distribution gives the following results:

(i) Coefficient of variation = 5

(ii) Standard deviation = 2

(iii) Karl Pearson's coefficient of skewness = 0.5

Find the mean and mode of the distribution.

Solution:

We are given C.V. = 5, σ = 2, Karl Pearson's coeff. of Sk = 0.5. We have to find mean and mode.

$$\text{C.V.} = \frac{\sigma}{\overline{X}} \times 100$$

$$5 = \frac{2}{\overline{X}} \times 100 \text{ or } 5\overline{X} = 200 \text{ or } \overline{X} = 40$$

$$\text{Coeff. of Sk} = \frac{\overline{X} - \text{Mode}}{\sigma} \quad 0.5 = \frac{40 - \text{Mode}}{2}$$

$$40 - \text{Mode} = 1 \text{ or } \text{Mode} = 39$$

Hence mean of the distribution is 40 and mode = 39.

Example 2:

By using the quartiles find a measure of skewness for the following distribution.

Annual Sales (Rs. 1000)	***No. of firms***
Less than 20	*30*
Less than 30	*225*
Less than 40	*465*
Less than 50	*580*
Less than 60	*634*
Less than 70	*644*
Less than 80	*650*
Less than 90	*665*
Less than 100	*680*

Solution:

Since we are given cumulative frequencies, we first determine simple frequencies and then apply the following formula based on quartiles.

$$Sk = L + \frac{Q_3 + Q_1 - 2\text{ Median}}{Q_3 - Q_1}$$

Calculation of Q_1, Q_3 and Median

Sales	f	c.f.
10–20	30	30
20–30	195	225
30–40	240	465
40–50	115	580
50–60	54	634
60–70	10	644
70–80	6	650
80–90	15	665
90–100	15	680

$$Q_1 = \text{Size of } \frac{N}{4}\text{th item} = \frac{680}{4} = 170\text{th item}$$

Q_1 lies in the class 20–30

$$Q_1 = L + \frac{N/4 - c.f.}{f} \times i \quad L = 20,\ N/4 = 170,$$

c.f. = 30, f = 195, i = 10

$$\therefore \quad Q_1 = 20 + \frac{170 - 30}{195} \times 10 = 20 + 7.18 = 27.18$$

$$Q_3 = \text{Size of } \frac{3N}{4}\text{th item} = \frac{3 \times 680}{4} = 510\text{th item}$$

Q_3 lies in the class 40–50

$$Q_3 = L + \frac{2N/4 - c.f.}{f} \times i \quad L = 40,\ 3N/4 = 510,$$

c.f. = 465, f = 115, i = 10

$$Q_3 = 40 + \frac{510 - 465}{115} \times 10 = 40 + 3.9 = 43.9$$

Median : Median = Size of $\frac{N}{2}$th item = $\frac{680}{2}$ = 340th item

Median lies in the class 30–40

$$\text{Med.} = L + \frac{N/2 - \text{c.f.}}{f} \times i$$

L = 30, N/2 = 340, c.f. = 225, f = 240, i = 10

$$\text{Med.} = 30 + \frac{340 - 225}{240} \times 10$$

$$= 30 + 4.79 = 34.79$$

$$\text{Coeff. of Sk} = \frac{43.9 + 27.18 - 2(34 - 79)}{43.9 - 27.18}$$

$$= \frac{71.08 - 69.58}{16.72} = \frac{1.5}{16.72} = 0.09.$$

Example 3:

From the following data calculate Karl Pearson's coefficient of skewness:

Marks more than	*0*	*10*	*20*	*30*	*40*	*50*	*60*	*70*	*80*
No. of students	*150*	*140*	*100*	*80*	*80*	*70*	*30*	*14*	*0*

Solution:

Calculation of Karl Pearson's Coefficient of Skewness

Marks	**m.p.** **m**	**No. of students** **f**	**(m – 45)/10** **d**	**fd**	**fd³**	**c.f.**
0–10	5	10	– 4	– 40	150	10
10–20	15	40	– 3	– 120	360	50
20–30	25	20	– 2	40	80	70
30–40	35	0	– 1	0	0	70
40–50	45	10	0	0	0	80
50–60	55	40	+ 1	+ 40	40	120
60–70	65	16	+ 2	+ 32	64	136
70–80	75	14	+ 3	42	125	150
		N = 150		Σ fd = – 86	Σ fd² = 830	

$$\text{Karl Pearson's Coeff. of Sk} = \frac{\overline{X} - \text{Mode}}{\sigma}$$

$$\overline{X} = A + \frac{\Sigma fd}{N} \times i = 45 - \frac{86}{150} \times 10 = 45 - 5.73 = 39.27$$

$$\sigma = \sqrt{\frac{\Sigma fd^2}{N} - \left(\frac{\Sigma fd}{N}\right)^2} \times i = \sqrt{\frac{830}{150} - \left(\frac{-86}{150}\right)^2} \times 10$$

$$= \sqrt{5.533 - .329} \times 10 = 2.281 \times 10 = 22.81.$$

For finding out mode we have to prepare grouping table and analysis table.

Grouping Table

Marks	**f** I	**II**	**III**	**IV**	**V**	**VI**
0–10	10	50		70		
10–20	40		60		60	
20–30	20	20				30
30–40	0		10	50		
40–50	10	50			66	
50–60	40		56			70
60–70	16	30				
70–80	14					

Analysis Table

	Class in which mode is expected to lie		
Col. No.	**10–20**	**20–30**	**50–60**
I	1		1
II	1		1
III	1	1	
IV	1	1	
V			1
VI			1
	4	2	4

This is a bimodal series and hence for calculating skewness the following formula shall be used:

$$Sk = \frac{3(\overline{X} - \text{Med})}{\sigma}$$

Med = size of N/2 = 150/2 = 75th item. It lies in the class 40–50

$$\text{Med} = L = \frac{N/2 - c.f.}{f} \times i$$

L = 40, N/2 = 75, c.f. = 70, f = 10, i = 10

$$\text{Med} = 40 + \frac{75 - 70}{10} \times 10 = 40 + 5 = 45$$

$$\text{Coeff. of Sk} = \frac{3(39.27 - 45)}{22.81} = \frac{-17.19}{22.81} = -0.754$$

Example 4:

Pearson's coefficient of skewness for a distribution is 0.4 and coefficient of variance is 30%. Its mode is 88. Find the mean and the median.

Solution:

We are given $Sk_p = 0.4$, mode = 88 and coeff. of variance 30%. We have to calculate mean and median

$$Sk = \frac{\text{Mean} - \text{Mode}}{\sigma}$$

Coeff. of variance is 30% or $\frac{\sigma}{\bar{x}} = 0.3$

$$= \frac{1 - \frac{\text{mode}}{\text{mean}}}{\frac{\text{S.D.}}{\text{Mean}}} = 1 - \frac{88}{\text{Mean}} = .4 \times .3 = .12$$

$$\frac{88}{\text{Mean}} = 1 - .12 = .88$$

$$88 \text{ mean} = 88 \text{ or mean} = \frac{88}{.88} = 100$$

$$\text{mode} = 3 \text{ med.} - 2\bar{x}$$

$$88 = 3 \text{ med.} - 2 \text{ (M) or } 3 \text{ med } 288 \text{ or median} = 96$$

Hence mean and median are 100 and 96 respectively.

Example 5:

The daily expenditure of 100 families is given below:

Daily Expenditure:	*0–20*	*20–40*	*40–60*	*60–80*	*80–100*
No. of families:	*13*	*?*	*27*	*?*	*16*

If the mode of the distribution is 44, calculate Karl Pearson's coefficient of skewness.

Solution:

Let the missing frequency for the class 20–40 be X. The frequency for the class 60–80 shall be 100 – (X + 56) = 44 – X.

Expenditure	f	c.f.
0–20	13	13
20–40	X	13 + X
40–60	27	40 + X
60–80	(44 – X)	84
80–100	16	100

$$Mo = L + \frac{\Delta_1}{\Delta_1 + \Delta_2} \times i$$

Since modal value is 44, it lies in the class 40–60,

$$Mo = 44,\ L = 40,\ \Delta_1 = 27 - X$$

$$\Delta_2 = 27 - (44 - X) = 27 - 44 + X = X - 17,\ i = 20$$

$$44 = 40 + \frac{27 - X}{27 - X + X - 17} \times 20$$

$$4 = \frac{27 - X}{10} \times 20 = (27 - X)^2$$

or $27 - X = 2$ or $X = 25$

Thus the frequency for the class 20–40 is X and the frequency of the class 60–80 is 44 – 25 = 19. Thus the completed frequency distribution is:

0–20	20–40	40–60	60–80	80–100
13	25	27	19	16

Calculation of Coefficient of Skewness

Daily Expenditure	m.p. m	f	(m – 50)/20 d	fd	fd^2
0–20	10	13	– 2	– 26	52
20–40	30	25	– 1	– 25	25
40–60	50	27	– 0	0	0
60–80	70	19	+ 1	+ 19	19
80–100	90	16	+ 2	+ 32	64
		N = 100		$\Sigma fd = 0$	$\Sigma fd^2 = 160$

$$\text{Coeff. of Sk} = \frac{\overline{X} - \text{Mode}}{\sigma}$$

$$\overline{X} = A + \frac{\Sigma fd}{N} \times i$$

$$= 50 + \frac{0}{100} \times 20 = 50$$

$$\sigma = \sqrt{\frac{\Sigma fd^2}{N} - \left(\frac{\Sigma fd}{N}\right)^2} \times i = \sqrt{\frac{160}{100} - \left(\frac{0}{100}\right)^2} \times 20$$

$$= \sqrt{16} \times 20$$

$$= 1.265 \times 20 = 25.3$$

$$\text{Coeff. of Sk} = \frac{50 - 44}{25.3} = \frac{6}{25.3} = 0.237$$

Example 6:

Find Karl Pearson's coefficient of skewness for the following distribution and comment on the result:

Class :	*3–7*	*8–12*	*13–17*	*18–22*	*23–27*	*28–32*
Frequency :	*108*	*580*	*175*	*80*	*32*	*18*

Solution:

First convert the distribution to exclusive form and then find skewness.

Calculation of Karl Pearson's Coefficient of Skewness

Class	**m.p. m**	**f**	**(m – 15)/5 d**	**fd**	**fd²**	**c.f.**
2.5–7.5	5	108	– 2	– 216	432	108
7.5–12.5	10	580	– 1	– 580	580	688
12.5–17.5	15	175	0	0	0	863
17.5–22.5	20	80	+ 1	+ 80	80	943
22.5–27.5	25	32	+ 2	+ 64	128	975
27.5–32.5	30	18	+ 3	+ 54	162	993
		N = 993		Σ fd = – 598	Σ fd² = 1382	

$$Sk_P = \frac{\text{Mean} - \text{Mode}}{\sigma}$$

Calculation of Mean : $\overline{X} = A + \frac{\Sigma fd}{N} \times i$

$$= 15 - \frac{598}{993} \times 5 = 15 - 3.01 = 11.99$$

Calculation of Mode : By inspection mode lies in the class 7.5–12.5

$$Mo = L + \frac{\Delta_1}{\Delta_1 + \Delta_2} \times i$$

$$L = 7.5,\ \Delta_1 = (580 - 108) = 472,$$

$$\Delta_2 = (580 - 175) = 405.,\ i = 5$$

$$Mo = 7.5 + \frac{472}{472 + 405} \times 5$$

$$= 7.5 + 2.69 = 10.19$$

Calculation of S.D. :

$$\sigma = \sqrt{\frac{\Sigma fd^2}{N} - \left(\frac{\Sigma fd}{N}\right)^2} \times i = \sqrt{\frac{1382}{993} - \left(\frac{-598}{993}\right)^2} \times 5$$

$$= \sqrt{1.392 - .363} \times 5 = 5.07$$

$$\text{Coeff. of Sk} = \frac{11.99 - 10.19}{5.07} = 0.355.$$

Example 7:

Calculate the first four moments about the mean from the following data. Also calculate the value of β_1 and β_2.

Marks	*No. of students*
0–10	5
10–20	12
20–30	18
30–40	40
40–50	15
50–60	7
60–70	3

Solution:

Calculation of Moments about Mean and β_1, β_2

Marks	f	m.p. m	(m – 35)/10 d	fd	fd²	fd³	fd⁴
0–10	5	5	– 3	– 15	45	– 135	405
10–20	12	15	– 2	– 24	48	– 96	192
20–30	15	25	– 1	– 18	18	– 18	18
30–40	40	35	0	0	0	0	0
40–50	15	45	+ 1	+ 15	15	+ 15	15
50–60	7	55	+ 2	+ 14	28	+ 56	112
60–70	3	65	+ 3	+ 9	27	+ 81	243
	N = 100			Σfd = 19	Σfd² = 181	Σ fd³ = 97	Σ fd⁴ = 985

$$\mu_1' = \frac{\Sigma fd}{N} \times i = \frac{-19}{10} \times 10 = -1.9$$

$$\mu_2' = \frac{\Sigma fd^2}{N} \times i^2 = \frac{181}{100} \times 100 = 181$$

$$\mu_3' = \frac{\Sigma fd^3}{N} \times i^3 = \frac{-97}{100} \times 1000 = -970$$

$$\mu_4' = \frac{\Sigma fd^4}{N} \times i^4 = \frac{985}{100} \times 10000 = 98500$$

$$\mu_2 = \mu_2' - (\mu_1')^2$$
$$= 181 - (-1.9)^2 = 181 - 3.61 = 177.39$$

$$\mu_3 = \mu_3' - 3\mu_1'\mu_2' + 2\mu_1'^3$$
$$= -970 - 3(-1.9)(181) + 2(-1.9)^3$$
$$= -970 + 1031.7 - 13.718 = 47.982$$

$$\mu_4 = \mu_4' - 4\mu_1'\mu_3' + 6\mu_1'^2\mu_2' - 3\mu_1'^4$$
$$= 98500 - 4(-1.9)(-970) + 6(-1.9)^2\, 181 - 3(-1.9)^4$$
$$= 98500 - 7372 + 3920.46 - 39.096 = 95009.364$$

$$\beta_1 = \frac{\mu_3^2}{\mu_2^3} = \frac{(47.982)^2}{(177.29)^3} = \frac{2302.27}{5581968.75} = 0.0004$$

$$\beta_2 = \frac{\mu_4}{\mu_2^2} = \frac{95009.364}{(177.39)^3} = \frac{95009.364}{31467.212} = 3.02$$

Example 8:

If it is given that $\sum fx' = -100$, $\sum fx'^2 = 400$, $\sum fx^2 = 400$, $\sum fx'^3 = -1000$, $\sum fx'^4 = 5000$ and $N = 10$, find the first four central moments.

Solution:

$$\mu_1' = \frac{\Sigma fx'}{N} = \frac{-100}{10} = -10;$$

$$\mu_2' = \frac{\Sigma fx'^2}{N} = \frac{400}{10} = 40;$$

$$\mu_3' = \frac{\Sigma fx'^3}{N} = \frac{-1000}{10} = -100;$$

$$\mu_4' = \frac{\Sigma fx'^4}{N} = \frac{5000}{10} = 500;$$

$$\mu_2 = \mu_2' - (\mu_1')^2 = 40 - (-10)^2 = 40 - 100 = -60$$

$$\mu_3 = \mu_3' - 3\mu_1'\mu_2' + 2\mu_1'^3$$

$$= -100 - 3(-10)(40) + 2(-10)^3$$

$$= -100 + 1200 - 2000 = -900$$

$$\mu_4 = \mu_4' - 4\mu_1'\mu_3' + 6\mu_1'^2\mu_2' - 3(\mu_1')^4$$

$$= 500 - 4(-10)(-100) + (-10)^2(40) - 3(-10)^4$$

$$= 500 - 4000 + 24000 - 30000 = -9500.$$

Example 9:

First three moments of a variable measured by point "2" are gradually 1, 16 and – 40. Prove that mean is 3, variance is 15 and $\mu_4 = -86$.

Solution:

We are given moments about arbitrary origin 2, *i.e.*,

$$\mu_1' = 1, \mu_2' = 16, \mu_3' = -40.$$

Mean of $\overline{X} = A + \mu_1' = 2 + 1 = 3$

$$\mu_2 = \mu_2' - \mu_1'^2 = 16 - (1)^2 = 15$$

$$\mu_3 = \mu_3' - 3\mu_1'\mu_2' + 2\mu_1'^3$$

$$= -40 - 3(1)(16) + 2(1)^3$$

$$= -40 - 48 + 2 = -86.$$

Example 10:

The first four moments of a distribution about X = 2 are – 2, 12, – 20 and 100. Calculate the moment about mean. Also calculate β_2 and show whether the distribution is leptokurtic or platykurtic.

Solution:

We are given $\mu_1' = -2$, $\mu_2' = 12$, $\mu_3' = -20$ and $\mu_4' = 100$

$$\mu_2 = \mu_2' - (\mu_1')^2 = 12 - (-2)^2 = 12 - 4 = 8$$

$$\mu_3 = \mu_3' - 3\mu_1'\mu_2' + 2\mu_1'^3$$

$$= -20 - 3(-2)(12) + 2(-2)^3$$

$$= -20 + 72 - 16 = 36$$

$$\mu_4 = \mu_4' - 4\mu_1'\mu_3' + 6\mu_1'^2\mu_2' - 3\mu_1'^4$$

$$= 100 - 4(-2)^2(-20) + 6(-2)^2(12) - 3(-2)^4$$

$$= 100 - 320 + 288 - 48 = 20$$

$$\beta_2 = \frac{\mu_4}{\mu_2^2} = \frac{20}{(8)^2} = 0.3125$$

Since the value of β_2 is less than 3 the distribution is platykurtic.

Example 11:

For a distribution Bowley's coefficient of skewness is – 0.56, $Q_1 = 16.4$ and median = 24.2. What is the coefficient of quartile deviation?

Solution:

We are given $Sk_B = -0.56$, $Q_1 = 16.4$, Med. = 24.2

$$Sk_B = \frac{Q_3 + Q_1 - 2\text{ Med.}}{Q_3 - Q_1}$$

$$-0.56(Q_3 - 16.4) = Q_3 + 16.4 - 2(24.2)$$

$$-0.56\,Q_3 + 9.184 = Q_3 + 16.4 - 48.4$$

$$-0.56\,Q_3 - Q_3 = 16.4 - 48.4 - 9.184$$

$$-1.56\,Q_3 = 41.184, \text{ or } Q_3 = 26.4$$

$$\text{Q.D.} = \frac{Q_3 - Q_1}{2} = \frac{26.4 - 16.4}{2} = 5$$

Example 12:

Given $Q_1 = 18$, $Q_3 = 25$, Mode = 21, Mean = 18, find the coefficient of skewness.

Solution:

$$\text{Coeff. of Sk} = \frac{Q_3 + Q_1 - 2\,\text{Med}}{Q_3 - Q_1}$$

We are given Q_1, Q_2 and mode. But we need median for calculating skewness. The value of median can be determined as follows:

$$\text{Mode} = 3\text{ Med.} - 2\ \overline{X}$$

$$21 = 3\text{ Med. } 2 \times 18$$

$$3\text{ Med.} = 57 \text{ of Median} = 19$$

$$\text{Coeff. of Sk} = \frac{25 + 18 - 2(19)}{25 - 18} = \frac{43 - 38}{7} = +0.714$$

Example 13:

From the information given below calculate Karl Pearson's coefficient of skewness

Measure	***Place A***	***Place B***
Mean	*256.5*	*240.8*
Median	*201.1*	*201.6*
S.D.	*215.0*	*181.0*

Solution:

We have to determine skewness. Since we are not given mode, we shall calculate mode by the formula :

$$\text{Mode} = 3\text{ median} - 2\text{ Mean}$$

Place A Median = 201, Mean = 256.5

$$\text{Mode} = 3\,(201) - 2\,(256.5) = 603 - 513 = 90$$

$$\text{Coeff. of Sk} = \frac{\text{Mean} - \text{Mode}}{\sigma} = \frac{256.5 - 90}{215} = 0.774$$

Place B Mode = 3 (201.6) – 2 (240.8)

$$= 604.8 - 481.6 = 123.2$$

$$\text{Coeff. of Sk} = \frac{\text{Mean} - \text{Mode}}{\sigma} = \frac{240.3 - 123.2}{181} = 0.649.$$

Example 14:

The standard deviation of symmetrical distribution is 3. What must be the value of the fourth moment about the mean in order that the distribution be mesokurtic?

(b) If the first four moments of distribution about the value 45 are equal to – 4, 22, – 117 and 560, determine the corresponding moments :

(i) about the mean, and (ii) about zero.

Solution:

(a) For a mesokurtic distribution $\beta_2 = 3$

$$\beta_2 = \frac{\mu_4}{\mu_2^2}$$

We are given $\sigma = 3$

$$\mu_2 = \sigma^2 = (3)^2 = 9,$$

$$\beta_2 = 3, \mu_2 = 9$$

$$3 = \frac{\mu_4}{9^2} \Rightarrow \mu_4 = 243$$

Thus the fourth moment about mean must be 243 in order that distribution be mesokurtic.

(b) We are given moments about an arbitrary origin 5.

Thus $\mu_1' = -4$, $\mu_2' = 22$, $\mu_3' = -117$, $\mu_4' = 560$.

Moments about Mean :

From these we can find out moments about mean from the following relationships :

$$\mu_2 = \mu_2' - (\mu_1')^2 \qquad \mu_1'\mu_2' + 2(\mu_1')^3$$

$$\mu_3 = \mu_3' - 32$$

$$\mu_4 = \mu_4' - 4\mu_1'\mu_3' + 6\mu_2'(\mu_1)^2 - 3(\mu_1')^4$$

Substituting the values,

$$\mu_2 = 22 - (-4)^2 = 22 - 16 = 6$$

$$\mu_3 = -117 - 3(-4)(22) + 2(-4)^3 = -117 + 264 - 128 = 19$$

$$\mu_4 = 560 - 4(-4)(117) + 6(22)(-4)^2 - 3(-4)^4$$

$$= 560 - 1872 + 2112 - 768 = 32$$

Thus the moments about mean are $\mu_1 = 0$, $\mu_2 = 6$, $\mu_3 = 19$ and $\mu_4 = 32$.

Moments about Zero :

Let the moments about zero be denoted by v_1, v_2, v_3 etc.

The first moment about zero *i.e.*, v_1 or mean

The second moment about zero *i.e.*, $v_3 = \mu_1 + (v_1)^2$

The third moment about zero *i.e.*, $v_2 = \mu_2 + 3v_1v_2 - 2v_1^2$

The fourth moment about zero *i.e.*, $v_4 = \mu_4 + 4v_1v_2 - 6(v_1)^2 v_2 + 3v_1^4$

Substituting these values, $v_1 = 5 + (-4) = 1$

$v_2 = 6 + (1)^2 = 7$

$v_3 = 19 + 3(1)(7) - 2(1)^2 = 19 + 21 - 2 = 38$

$v_4 = 32 + 4(1)(38) - 6(1)^2(7) + 3(1)^4$

$= 32 + 152 - 42 + 3 = 145$

The moments about zero are $v_1 = 1$, $v_2 = 7$, $v_3 = 38$, $v_4 = 145$.

Example 15:

The following data are given to an economist for the purpose of economic analysis. The data refer to the length of a certain type of batteries.

$n = 100$, $\Sigma fd = 50$, $\Sigma fd^2 = 1970$, $\Sigma fd^3 = 2948$ and

$\Sigma fd^4 = 86{,}752$, in which $d = (X - 48)$.

Do you think that the distribution is platykurtic?

Solution:

For finding out whether the distribution is platykuric or not, we have to calculate β_2

$$\beta_2 = \frac{\mu_4}{\mu_2^2}$$

$$\mu_1' = \frac{\Sigma fd}{N} = \frac{50}{100} = 0.5$$

$$\mu_2' = \frac{\Sigma fd^2}{N} = \frac{1970}{100} = 19.7$$

$$\mu_3' = \frac{\Sigma fd^3}{N} = \frac{2948}{100} = 29.48$$

$$\mu_4' = \frac{\Sigma fd^4}{N} = \frac{86752}{100} = 867.52$$

$\mu_2 = \mu_2' - (\mu_1')^2 = 19.7 - (.5)^2 = 19.45$

$\mu_3 = \mu_3' - 3\mu_1'\mu_2' + 2(\mu_1')^3$

$= 29.48 - 3(.5)(19.7) + 2(.5)^3$

$= 29.48 - 29.55 + 0.25 = 0.18$

$\mu_4 = \mu_4' - 4\mu_1'\mu_3' = 6\mu_1'^2\mu_2' - 3(\mu_1')^4$

$= 867.52 - 4(.5)(29.48) + 6(19.7)(.5)^2 - 3(.5)^4$

$$= 867.52 - 58.96 + 29.25 - .1875 = 837.9225$$

$$\beta_2 = \frac{837.9225}{(19.45)^2} = \frac{837.9225}{378.3025} = +2.214.$$

Since the value of β_2 is less than 3, the distribution is platykurtic.

Example 16:

From the following data of the wages of 50 workers of a factory compute the first four moments about mean and also the value of β_1, β_2. Comment on the results.

Weekly wages (Rs.)	*No. of workers*	*Weekly wages (Rs.)*	*No. of workers*
100–120	*1*	*180–200*	*12*
120–140	*3*	*200–220*	*4*
140–160	*7*	*220–240*	*3*
160–180	*20*		

Solution:

Calculation of Moments and β_1, β_2

Wages (Rs.)	m.p. m	f	(m – 170)/10 d	fd	fd²	fd³	fd⁴
100–120	110	1	– 3	– 3	9	– 27	81
120–140	130	3	– 2	– 6	12	– 24	48
140–160	150	7	– 1	– 7	7	– 7	7
160–180	170	20	0	0	0	0	0
180–200	190	12	+ 1	+ 12	12	+ 12	12
200–220	210	4	+ 2	+ 8	16	+ 32	64
220–240	230	3	+ 3	+ 9	27	+ 81	243
		N = 50		Σ fd = 13	Σ fd² = 83	Σ fd = 67	Σ fd⁴ = 455

Moments about Arbitrary Origin 170

$$\mu_1' = \frac{\Sigma fd}{N} \times i = \frac{13}{50} \times 10 = 2.6.$$

$$\mu_2' = \frac{\Sigma fd^2}{N} \times i^2 = \frac{83}{50} \times 100 = 166.$$

$$\mu_3' = \frac{\Sigma fd^3}{N} \times i^3 = \frac{67}{50} \times 1000 = 1340.$$

$$\mu_4' = \frac{\Sigma fd^4}{N} \times i^4 = \frac{455}{50} \times 10000 = 91000.$$

Moments about Mean

$$\mu_2 = \mu_2' = (\mu_1)^2 = 166 - (2.6)^2 = 166 - 6.76 = 159.24$$

$$\mu_3 = \mu_3' - 3\mu_1'\mu_2' + 2\mu_1'^3 = 1340 - 3\,(2.6)\,(166) + 2\,(2.6)^3$$

$$= 1340 - 1294.8 + 35.152 = 80.352$$

$$\mu_4 = \mu_4' - 4\mu_1'\mu_3' + 6\,(\mu_1')^2\,(\mu_2') - 3\,(\mu_1')^4$$

$$= 91000 - 4\,(2.6)\,(1340) + 6\,(2.6)^2\,(166) - 3\,(2.6)^4$$

$$= 91000 - 13936 + 6732.96 - 137.09 = 83659.87$$

$$\beta_1 = \frac{\mu_3^2}{\mu_2^2} = \frac{(80.325)^2}{(159.24)^3} = \frac{6456.444}{40378908.81} = 0.0016$$

Since the value of b_1 is very close to zero, the distribution is almost symmetrical.

$$\beta_2 = \frac{\mu_4}{\mu_2^2} = \frac{83659.87}{(159.24)^2} = \frac{83659.87}{25357.38} = 3.3$$

Since the value of β_2 is more than 3 the distribution is platykurtic.

Example 17:

You are given the following values of moments :

$\mu_2 = 44.553, \qquad \mu_3 = -9.774, \qquad \mu_4 = 5508.567.$

Find the corrected values of each one of these, taking into account the class interval which is 3.

Solution:

$$\mu_2 \text{ (corrected)} = \mu_2 \text{ (uncorrected)} - \frac{i^2}{12}$$

$$= 43.353 - \frac{(3)^2}{12} = 43.353 - 0.75 = 42.603$$

$$\mu_4 \text{ (corrected)} = \mu_4 \text{ (uncorrected)} - \frac{1}{12}\, i^2\, \mu_2 \text{ (uncorrected)} + \frac{7i^2}{240}$$

$$= 5508.567 - \frac{1}{2}\,(3)^2\,(43.353) + \frac{7(3)^4}{250}$$

$$= 5508.567 - 195.0885 + 2.3625 = 5315.841.$$

Example 18:

Calculate β_1 and β_2 from the following distribution and interpret the results.

Age	*Frequency*	*Age*	*Frequency*
25–30	*2*	*45–50*	*25*
30–35	*8*	*50–55*	*16*
35–40	*18*	*55–60*	*7*
40–45	*27*	*60–65*	*2*

Solution:

Calculation of β_1 and β_2

Age	m.p. m	f	(m – 42.5)/5 d	fd	fd²	fd³	fd⁴
25–30	27.5	2	– 3	– 6	18	– 54	162
30–35	32.5	8	– 2	– 16	32	– 64	128
35–40	37.5	18	– 1	– 18	18	– 18	18
40–45	42.5	27	0	0	0	0	0
45–50	47.5	25	+ 1	+ 25	25	+ 25	25
50–55	52.5	16	+ 2	+ 32	64	+ 128	256
55–60	57.5	7	+ 3	+ 21	63	+ 189	567
60–65	62.5	2	+ 4	+ 8	32	+ 128	512
		N = 105		Σ fd = 46	Σ fd² = 252	Σ fd³ = 334	Σ fd⁴ = 1668

$$\mu_1' = \frac{\Sigma fd}{N} = \frac{46}{105} = 0.438;$$

$$\mu_2' = \frac{\Sigma fd^2}{N} = \frac{252}{105} = 2.4$$

$$\mu_3' = \frac{\Sigma fd^3}{N} = \frac{334}{105} = 3.181$$

$$\mu_4' = \frac{\Sigma fd^4}{N} = \frac{1668}{105} = 15.886$$

$$\mu_2 = \mu_2' - (\mu_1')^2 = 2.4 - (.438)^2 = 2.4 - .192 = 2.238$$

$$\mu_3 = \mu_3' - 3\mu_1'\mu_2' + 2\,(\mu_1')^3$$

$$= 3.181 - 3\,(.438)\,(2.4) - 2\,(.438)^3$$

$$= 3.181 - 3.1536 + .1681 = 0.1955$$

$$\mu_4 = \mu_4' - 4\mu_1'\mu_3' + 6\mu_1'^2\mu_2' - 3(\mu_1')^4$$

$$= 15.886 - 4\,(.438)\,(3.181) + 6\,(.438)^2\,(2.4) - 3\,(.439)^4$$

$$= 15.886 - 5.573 + 2.763 - .110 = 12.966$$

$$\beta_1 = \frac{\mu_3^2}{\mu_2^3} = \frac{(.1955)^3}{(2.238)^3} = 0.0034;$$

$$\beta_2 = \frac{\mu^4}{\mu_2^3} = \frac{12.966}{(2.238)^2} = 2.59$$

For finding the values of β_1 and β_2 it is not necessary to multiply μ_1', μ_2', μ_3' by i, i^2, i^3 because β_1 and β_2 are independent of change of scale and origin.

Example 19:

A frequency distribution showed the following measures of iocation:

Mean = 45, Median = 48, Coefficient of skewness = – 0.4.

Estimate its standard deviation with the help of the above information.

Solution:

$$\text{Coeff. of Sk} = \frac{3(\overline{X} - \text{Med.})}{\sigma} \quad \overline{X} = 45,\ \text{Med.} = 48,$$

Coeff. of Sk = – 0.4

Substituting the values $-\,0.4 = \dfrac{3(45 - 48)}{\sigma} = .4\sigma = -\,9 \text{ or } \sigma = 2.25.$

Example 20:

Calculate Karl Pearson's coefficient of skewness from the following data:

Marks more than :	*5*	*15*	*25*	*35*	*45*	*55*	*65*	*75*	*85*
No. of students :	*120*	*105*	*96*	*85*	*72*	*58*	*42*	*12*	*0*

Solution:

First find simple frequencies by taking classes as 5–15, 15–25 etc. then calculate coefficient of skewness.

Calculation of Karl Pearson's Coefficient of Skewness

Marks	m	f	(m – 50)/10 d	fd	fd²	c.f.
5–15	10	15	– 4	– 60	240	15
15–25	20	9	– 3	– 27	81	24
25–35	30	11	– 2	– 22	44	35
35–45	40	13	– 1	– 13	13	48
45–55	50	14	0	0	0	62
55–65	60	16	+ 1	+ 16	16	78
65–75	70	30	+ 2	+ 60	120	108
75–85	80	12	+ 3	+ 35	108	120
		N = 120		Σ fd = – 10	Σ fd² = 622	

Since the distribution is lightly skewed and the value of mode is undefined we apply the formula based on relationship between mean, median

$$Sk = \frac{3(\overline{X} - \text{Med.})}{\sigma}$$

Calculation of Mean:

$$\overline{X} = A + \frac{\Sigma fd}{N} \times i$$

$$= 50 - \frac{10}{120} \times 10 = 50 - .83 = 49.17$$

Calculation Median :

$$\text{Med.} = \text{Size of } \frac{N}{2}\text{th item} = 60\text{th item}$$

Median lies in the class 45–55

$$\text{Med.} = L + \frac{N/2 - \text{c.t.}}{f} \times i$$

L = 45, N/2 = 60,

c.f. = 48, f = 14, i = 10

$$\text{Med.} = 45 + \frac{60 - 48}{14} \times 10 = 45 + 8.57 = 53.57$$

Calculation of S.D.

$$\sigma = \sqrt{\frac{\Sigma fd^2}{N} - \left(\frac{\Sigma fd}{N}\right)^2} \times i = \sqrt{\frac{622}{120} - \left(\frac{-10}{120}\right)^2} \times 10$$

$$= \sqrt{5.183 - .007} \times 10 = 2.275 \times 10 = 22.75$$

$$Sk = \frac{3(49.17 - 53.57)}{2275} = -\frac{13.2}{22.75} = -0.58$$

Example 21:

Calculate coeff. of Q.D. and Bowley's coeff. of skewness from the data given below:

Profits (Rs. Lakhs) :	*Less than 10*	*20*	*30*	*40*	*50*	*60*	*70*
No. of Cos.	*8*	*20*	*40*	*50*	*56*	*59*	*60*

Solution:

Calculation of Q.D. and Bowley's Coefficient of Skewness

Profits (Rs. lakhs)	f	c.f.
0–10	8	8
10–20	12	20
20–30	20	40
30–40	10	50
40–50	6	56
50–60	3	59
60–70	1	60
	N = 60	

$$Q.D. = \frac{Q_3 - Q_1}{2}$$

Q_1 = Size of $\frac{N}{4}$th item = $\frac{60}{4}$ = 15th item.

Q_1 lies in the class 10–20

$$Q_1 = L + \frac{N/4 - c.f.}{f} \times = 10 + \frac{15 - 8}{12} \times 10 = 10 + 5.833 = 15.833$$

Q_3 = Size of $\frac{3N}{4}$th item

$$= \frac{3 \times 60}{4} = 45\text{th item. Hence } Q_3 \text{ lies in the class 30–40}$$

$$Q_3 = L + \frac{3N/4 - c.f}{f} \times i = 30 + \frac{45 - 40}{10} \times 10 = 35$$

$$Q.D. = \frac{35 - 15.833}{2} = 9.584$$

$$Sk_B = \frac{Q_3 + Q_1 - 2\,Med}{Q_3 - Q_1}$$

$$Med. = \text{Size of } \frac{N}{2} \text{th item} = \frac{60}{2}$$

$$= 30\text{th item. Hence it lies in the class 20–30}$$

$$Med. = L + \frac{N/2 - c.f.}{f} \times i = 20 + \frac{30 - 20}{12} \times 10 = 20 + 8.33 = 28.33$$

$$Sk_B = \frac{35 + 15.833 - 2(29.33)}{35 - 15.833} = \frac{-5.827}{19.167} = -0.304.$$

Example 22:

Calculate coefficient of skewness by Karl Pearson's method and the values of β_1 and β_2 from the following data :

Profits (Rs. Lakhs) :	*10–20*	*20–30*	*30–40*	*40–50*	*50–60*
No. of Companies :	*18*	*20*	*30*	*22*	*10*

Solution:

Calculation of Karl Pearson's Coefficient of Skewness, β_1 and β_2

Profits (Rs. lakhs)	m.p. m	No. of cos. f	(m – 35)/10 ds	fd	fd²	fd³	fd⁴
10–20	15	18	– 2	– 36	72	– 144	288
20–30	25	20	– 1	– 20	20	– 20	20
30–40	35	30	0	0	0	0	0
40–50	45	22	+ 1	+ 22	22	+ 22	22
50–60	55	10	+ 2	+ 20	40	+ 80	160
		N = 100		$\Sigma fd = -14$	$\Sigma fd^2 = 154$	$\Sigma fd^3 = -62$	$\Sigma fd^4 = 490$

$$Sk_P = \frac{Mean - Mode}{\sigma}$$

Mean : $\overline{X} = A + \dfrac{\Sigma fd}{N} \times i$

$= 35 - \dfrac{14}{100} \times 10 = 35 - 1.4 = 33.6$

Mode: $\text{Mode} = L + \dfrac{\Delta_1}{\Delta_1 + \Delta_2} \times i$

By inspection, mode lies in the class 30–40

$L = 30, \Delta_1 = 30 - 20 = 10,$

$\Delta_2 = 30 - 22 = 8, i = 10$

$\text{Mode} = 30 + \dfrac{10}{10 + 8} \times 10 = 30 = 30 + 5.56 = 35.56.$

S.D. : $\sigma = \sqrt{\dfrac{\Sigma fd^2}{N} - \left(\dfrac{\Sigma fd}{N}\right)^2} \times i = \sqrt{\dfrac{154}{100} - \left(\dfrac{-14}{100}\right)^2} \times 10$

$= \sqrt{1.54 - 0.0196} \times 10 = 1.233 \times 10 = 12.33$

$Sk_P = \dfrac{33.6 - 35.56}{12.33} = \dfrac{-1.96}{12.33} = -0.159.$

Calculation of β_1 :

$\mu_1' = \dfrac{\Sigma fd}{N} = \dfrac{-14}{100} = -.14;$

$\mu_2' = \dfrac{\Sigma fd^2}{N} = \dfrac{154}{100} = 1.54$

$\mu_3' = \dfrac{\Sigma fd^3}{N} = \dfrac{-62}{100} = -.62;$

$\mu_4' = \dfrac{\Sigma fd^4}{N} = \dfrac{490}{100} = 4.9$

$\mu_2 = \mu_2' - (\mu_1')^2 = 1.54 - (-.14)^2 = 1.54 - .0196 = 1.5204$

$\mu_3 = \mu_3' - 3\mu_1'\mu_2' + 2\mu_1'^3 = .62 + .6468 - .0055 = .0213$

$\mu_4 = \mu_4' - 4\mu_1'\mu_3' + 6\mu_1'^2\mu_3' - 3\mu_1'^4$

$= 4.9 - 4(-.14)(-.62) + 6(-.14)^2(1.54) - 3(-.14)^4$

$= 4.9 - .3472 + .1811 + .0024 = 4.732$

$\sqrt{\beta_1} = \dfrac{\mu_3}{\sqrt{\mu_2^3}} - \dfrac{.0213}{\sqrt{(1.5204)^2}} = \dfrac{.0213}{1.8747} = 0.0114$

$$\beta_2 = \frac{\mu_4}{\sqrt{\mu_2^2}} - \frac{7.732}{\sqrt{(1.5204)^2}} = \frac{4.732}{2.312} = 2.047.$$

Note : β_1 and β_2 are independent of change of scale and origin and hence to simplify calculations there is no need to multiply μ_1', μ_2', μ_3' by i, i^2, i^3 etc.

Example 23:

From the following data of age of employees, calculate coefficient of skewness and comment on the result:

Age below (yrs) :	*25*	*30*	*35*	*40*	*45*	*50*	*55*
No. of employees:	*8*	*20*	*40*	*65*	*80*	*92*	*100*

Solution:

We are given age below 25, 30 etc. We first find simple frequencies taking first class to be 20–25.

Calculation of Karl Pearson's Coefficient of Skewness

Age	**m.p.**	**f**	**(m – 37.5)/5 d**	**fd**	**fd²**
20–25	22.5	8	– 3	– 24	72
25–30	27.5	12	– 2	– 24	48
30–35	32.5	20	– 1	– 20	20
35–40	37.5	25	0	0	0
40–45	42.5	15	+ 1	+ 15	15
45–50	47.5	12	+ 2	+ 24	48
50–55	52.5	8	+ 3	+ 24	72
		N = 100		Σ fd = – 5	Σ fd² = 275

$$\text{Coeff. of Sk} = \frac{\overline{X} - \text{Mode}}{\sigma}$$

Calculation of Mean : $\overline{X} = A = \frac{\Sigma fd}{N} \times i$

$$A = 37.5,\ \Sigma fd = -5,\ N = 100,\ i = 5$$

$$\overline{X} = 37.5 - \frac{5}{100} \times 5 = 37.5 - .25 = 37.25$$

Calculation of Mode : By inspection mode lies in the class 35–40

$$Mo = L + \frac{\Delta_1}{\Delta_1 + \Delta_2} \times i \quad L = 35, \Delta_1 = 25 - 20 = 5,$$

$$\Delta_2 = 25 - 15 = 10, i = 5$$

$$Mo = 35 + \frac{5}{5+10} \times 5 = 35 + 1.67 = 36.67$$

Calculation of S.D.

$$\sigma = \sqrt{\frac{\Sigma fd^2}{N} - \left(\frac{\Sigma fd}{N}\right)^2} \times i = \sqrt{\frac{275}{100} - \left(\frac{5}{100}\right)} \times 5$$

$$= \sqrt{2.75 - 0.0025} \times 5 = 8.29$$

$$\text{Coeff. of Sk} = \frac{37.25 - 36.67}{8.29} = 0.07.$$

LIST OF FORMULAES

Skewness

(i) Absolute Skewness = Mean – Mode (Karl Pearson's)

(ii) Absolute Skewness = $Q_3 + Q_1 - 2$ Med. (Bowley's)

(iii) $Sk_P = \frac{\text{Mean} - \text{Mode}}{\sigma}$ $\qquad Sk_P = \frac{3(\text{Mean} - \text{Median})}{\sigma}$

(iv) $Sk_B = \frac{Q_3 + Q_1 - 2\text{ Med.}}{Q_3 - Q_1}$ $\qquad Sk_K = \frac{P_{10} + P_{90} - \text{Med.}}{P_{90} - P_{10}}$

Moments

(i) Moments about mean

$$\mu_1 = \frac{\Sigma (X - \overline{X})}{N} = 0 \qquad \mu_3 = \frac{\Sigma (X - \overline{X})^3}{N}$$

$$\mu_2 = \frac{\Sigma (X - \overline{X})^2}{N} \qquad \mu_4 = \frac{\Sigma (X - \overline{X})^4}{N}$$

(ii) In a frequency distribution

$$\mu_1 = \frac{\Sigma f (X - \overline{X})}{N} \qquad \mu_2 = \frac{\Sigma f (X - \overline{X})^2}{N} \text{ etc.}$$

(iii) Moments about arbitrary origin

$$\mu_1' = \frac{\Sigma (X - A)}{N}; \qquad \mu_3' = \frac{\Sigma (X - A)^3}{N}$$

$$\mu_2' = \frac{\Sigma (X - A)^2}{N}; \qquad \mu_4' = \frac{\Sigma (X - A)^4}{N}$$

(iv) In a frequency distribution

$$\mu_1' = \frac{\Sigma f\,(X - A)}{N} \quad \text{or} \quad \mu_1' = \frac{\Sigma fd}{N} \times i$$

$$\mu_2' = \frac{\Sigma f\,(X - A)^2}{N} \quad \text{or} \quad \mu_2' = \frac{\Sigma fd^2}{N} \times i^3$$

Kurtosis

(i) $$\beta_1 = \frac{\mu_3^2}{\mu_2^3} \quad \text{or} \quad \gamma_1 = \frac{\mu_3}{\mu_3^{3/2}}$$

(ii) $$\beta_2 = \frac{\mu_4}{\mu_2^2} \quad \text{or} \quad \gamma_2 = \beta_2 - 3$$

EXERCISES

1. Find out Karl Pearson's coefficient of skewness from the following table :

Wages (Rs.) :	60–70	70–80	80–90	90–100	100–110	110–120	120–130	130–140
No. of persons :	12	18	35	42	50	45	20	8

[– 0.331]

2. Calculate the first four moments about the mean for the following data and comment on the natural of the distribution:

X :	1	2	3	4	5	6	7	8	9
Y :	1	6	13	25	30	22	9	5	2

[$\beta_1 = 0.048$, $\beta_2 = 2.97$]

3. With the help of a suitable diagram, explain skewness and kurtosis of a distribution.
4. The first two moments of a distribution about the value 1 are 2 and 25 respectively. Find the mean and standard deviation of the distribution.
5. Define skewness. What are the various measures of skewness?
6. What is kurtosis? How is it measured?
7. By using the quartiles, find a measure of skewness for the following distribution:

Annual Sales (Rs. 000)	*No. of firms*
Less than 20	30
Less than 30	225
Less than 40	465

Less than 50	580
Less than 60	634
Less than 70	644
Less than 80	650
Less than 90	665
Less than 100	680

The standard deviation of a symmetrical distribution is given to be 5. What must be the value of fourth moment about the mean in order that the distribution is mesokurtik?

8. Calculate the first four moments and find the value of β_1 and β_2 from the data given below:

Marks :	0–20	20–40	40–60	60–80	80–100
No. of students :	8	12	20	6	4

[$b_1 = 0.0486$; $b_2 = 2.563$].

9. From the following data, compute quartile deviation and the coefficient of skewness:

Size :	5–7	8–10	11–13	14–16	17–19
Frequency :	14	24	38	20	4

[Q.D. = 1.773, $Sk_B = 0.111$]

10. (a) Calculate the first four moments about the mean and also the value of b_2 from the following data:

m :	0	1	2	3	4	5	6	7	8
f :	1	8	28	156	170	56	28	8	1

[$\mu_2 = 1.294$; $\mu_3 = 0.642$; $\mu_4 = 6.582$; $\beta_2 = 3.931$]

11. What is kurtosis? How the measures of kurtosis help in understanding a frequency distribution?

12. (a) Define skewness and describe briefly the various tests of skewness.
    (b) Define kurtosis. If $\beta_1 = +1$, $\beta_2 = 4$ and variance = 9, find the values of μ_3 and μ_4 and comment on the nature of the distribution.
    (c) What do you mean by 'skewness'? How is skewness different from kurtosis?

13. (a) Define moments. "A frequency distribution can be described almost completely by the first four moments and two measures based on moments". Examine the statement.

(b) Indicate whether two distributions with the same means, standard deviations and coefficients of skewness must have same peakedness.

14. (a) Calculate measure of skewness based on quartiles from the following data :

Mid value :	115	125	135	145	155	165	175	185	195
Frequency :	6	25	48	72	116	60	38	22	3

(b) For a normal distribution, the first moment about origin is 35 and the second moment about 35 is 10. Find the first four central moments.

15. Find the measure of kurtosis for the following distribution:

Class :	45–52	52–59	59–66	66–73	73–80	80–87	87–94
Frequency :	4	9	12	4	3	2	1

16. Obtain Pearson's coefficient of skewness from the following data using the empirical relationship among arithmetic mean, median and mode.

Marks	*No. of students*	*Marks*	*No. of students*
0–10	10	45–50	10
10–20	40	50–60	40
20–30	20	60–70	16
30–40	0	70–80	14.

17. Calculate Karl Pearson's coefficient of skewness from the following data:

Measurement :	0–10	10–20	20–30	30–40	40–50	50–60	60–70
Frequency :	10	12	18	25	16	14	5

18. Tick the correct answer:

(a) When coefficient of skewness is zero, the distribution is:

(i) J-shaped (ii) U-shaped

(iii) Symmetrical (iv) L-shaped

(v) None of these.

(b) When $\beta_2 < 3$, the distribution is :

(i) Leptokurtic (ii) Platykurtic

(iii) Mesokurtic (iv) None of these

(c) The second moment about mean is :

(i) $\frac{\Sigma(X + x)^3}{N}$ (ii) $\frac{\Sigma(X - x)^3}{N}$

(iii) $\frac{\Sigma(X + x)^2}{N}$ (iv) $\frac{\Sigma(X - x)^2}{N}$

(v) None of these

(d) In a negative skewed distribution :

(i) mode < median < mean

(ii) mode < median > mean

(iii) median > mode > mean

(iv) mean > median > mode

(v) mode < median > mean

(e) γ_2 is :

(i) $\beta_2 + 3$ (ii) $\beta_1 - 3$

(iii) $\beta + 2$ (iv) None of these

(f) Mention the correct answer with reasoning in one line :

If a frequency distribution is positively skewed, the mean of the distribution is :

(i) equal to (ii) less than (iii) greater than

(iv) not related to Bowley's coefficient of skewness for any skewed distribution.

(g) Mention the correct answer with reasoning in one line:

If a frequency distribution is positively skewed, the mean of the distribution is :

(i) Greater than the mode (ii) Less than the mode

(iii) Equal to the mode (iv) None of these

(h) When coefficient of skewness is negative :

(i) $Q_3 + Q_1 = 2Q$ (ii) $Q_3 + Q_1 < 2Q_2$

(iii) $Q_3 + Q_1 > 2Q_2$ (iv) $Q_3 + Q_1 < 2Q_1$

(v) None of these

Ans. a. (iii) b. (ii) c. (iv) d. (v) e. (iii) f. (iv) g. (v) h. (i)

19. From the following data, compute quartiles and find the coefficient of skewness :

Income (Rs.)	*No. of persons*	*Income (Rs.)*	*No. of persons*
Below 200	25	600–800	75
200–400	40	800–1,000	20
400–600	80	above 1,000	16

20. (a) Calculate Bowley's coefficient of skewness from the following data :

Profits (Rs. Crores) :	10–20	20–30	30–40	40–50	50–60
No. of companies :	15	20	30	10	5

(b) Explain negative and positive skewness. Calculate quartile deviation and coefficient of skewness from the following :

Median : 18.8 inches, Q_1 = 14.6 inches, Q_3 = 25.2 inches.

21. When is skewness present in a series.

22. (i) Comment whether the following statements are true or false :

(a) Moments are useful to measure kurtosis?

(b) B_2 is a measure of skewness.

(ii) Fill in the blanks:

Skewness is positive when mean is ..... than mode.

(iii) Define Pearson's measure of skewness.

(iv) Explain the terms skewness and kurtosis. Give the different measures of skewness.

(v) What is skewness? What are the different measures of skewness?

23. Indicate whether the following statements are True of False :

(i) In a symmetrical distribution, mean = median = mode T/F

(ii) In a skewed distribution, quartiles are equidistant from the median. T/F

(iii) Skewness cannot be calculated in a bimodal distribution. T/F

(iv) β_2 is a measure of Kurtosis. T/F

(v) Moments about an arbitrary origin are called Central moments. T/F

(vi) Sheppard's corrections to eliminate grouping error must be made in every frequency distribution. T/F

(vii) The old moments in a symmetrical distribution are zero. T/F

Ans. : (i) T (ii) F (iii) F (iv) T (v) F (vi) F (viii) T

24. Calculate first four moments from the following data :

X:	0	1	2	3	4	5	6	7	8
f:	5	10	15	20	25	20	15	10	5

Also calculate the values of β_1 and β_2 and comment on the nature of distribution.

[$\beta_1 = 0, \beta_2 = 2.35$]

(b) It is given that $\Sigma fx' = -100$, $\Sigma fx'^2 = -400$, $\Sigma fx'^3 = 1{,}000$, $\Sigma fx'^4 = 5{,}000$ and N = 10. Find the first four central moments.

25. The following table gives the distribution of monthly wages of 500 *Workers in a factory :*

wages (Rs. hundrd) :	15–20	20–25	25–30	30–35	35–40	40–45
No. of workers :	1025	145	220	70	30	

Compute Karl Pearson's and Bowley's coefficient of skewness. Interpret your values.

26. Compute Karl Pearson's cefficient of skewness from the following data:

Weekly Sales (Lakh Rs.) :	10–12	12–14	14–16	16–18	18–20	20–22	22–24	24–26
No. of Computers :	12	18	35	42	50	45	30	8

Comment on the value obtained.

27. (a) Define kurtosis. For a distribution, the second, third and fourth central moments are 50, 100 and 6,600 respectively. Calculate a measure of kurtosis for this distribution.

28. (a) The standard deviation of symmetrical distribution in 3. What must be the value of the 4th moment almost the mean in order that the distribution be mesokurtic?

(b) Distinguish between Karl Pearson's and Bowley's measure of skewness. Which one of these would you prefer and why?

29. Explain briefly the different methods of measuring skewness.

30. Fill in the blanks :

(i) If $Q_3 = 30$, $Q_2 = 20$, Med. = 25. Coeff. of Sk. shall be ....

(ii) If $\overline{X} = 50$, Mode = 48, $\sigma = 20$, the coefficient of skewness shall be ....

(iii) If $\beta_2 = 3$, the distribution is called .....

If $\beta_2 < 3$, the distribution is .....

If $\beta < 3$, the distribution is ... make dots.

(iv) If Coeff. of Sk. = 3.8, median = 35, σ = 12, the mean shall be

(v) In a positively skewed distribution. ... > ... > ... > ... .

(vi) In a symmetrical distribution, the coefficient of skewness is ...

(vii) The limits for Bowley's coefficient of skewness are ...

(viii) In a symmetrical distribution, the distance between the ... and the ... about the distance between the ... and the ...

Ans. : (i) 0 (ii) 0.1 (iii) mesokurtic, leptokurtic, platykurtic, (iv) 108.2 (v) mean < median < mode. (vi) zero (vii) ± Ans. for (viii).

31. (a) Define moments. Also state the relationship up to the 4th order between the moments about men and the moments about the assumed mean.

(b) With the help of diagrams, show the various types of skewness and kurtosis, and also indicate their names.

32. (a) Define moments. How are they useful in analysing the different aspects of frequency distribution?

(b) What are moments? Explain the procedure of calculating the first four moments about the mean.

33. What is skewness? What are the tests of skewness? Draw different sketches to indicate different types of skewness and locate roughly the relative positions of mean, median and mode in each case.

34. (a) Giving illustrations, differentiate between moments and kurtosis. Explain their importance.

(b) Explain the term skewness as applied to a frequency distribution and describe various measures of skewness known to you.

35. (a) How would you analyse a frequency distribution through the method of moments?

(b) Distinguish between dispersion and skewness and point out the various methods of measuring skewness.

36. How does 'skewness' differ from dispersion? Elucidate the relative measures of skewness.

37. Find the moment coefficient of kurtosis from the following information:

4th moment = 189.37

2nd moment = 8.52

38. The following information was obtained from records of a factor relating to the wages :

Arithmetic mean	= Rs. 56.8
Median	= Rs. 59.5
Standard deviation	= Rs. 12.4

Give as much information as you can about the distribution of wages.

[C.V. = 21.8%; Sk. = 0.653]

39. (i) Calculate mode and median from the following information :

Karl Pearson's coeff. of Sk. = + 0.32

Standard deviation = 6.5

Arithmetic mean = 29.5

[Mode = 26.52, Med. = 28.57]

(ii) The first four moments of distribution about the value 4 are – 1.5, 17, – 30 and 108. Find : (i) Coefficient of variation (ii) β_1 (iii) First three moments about origin (zero).

[C.V. = 172, β_1 = 0.49]

40. Differentiate between Bowley's measure and Karl Pearson's measure of skewness.

41. What do you understand by skewness Distinguish clearly by giving figure between positive and negative skewness.

42. Calculate Bowely's coefficient of skewness from the data given below :

Mid-point :	35	45	55	65	75	85	95
Frequency :	1	3	11	21	43	32	9

[– 0.036]

43. Calculate Karl Pearson's coefficient of skewness from the data given below :

Variable	*Frequency*	*Variable*	*Frequency*
140–150	4	180–190	6
150–160	11	190–200	2
160–170	14	200–210	1
170–180	12		

[0.147].

44. Explain the third and fourth central moments in terms of the first four moments about the origin.

45. The following table gives the distribution of monthly income of workers in a factory :

Monthly income Rs.	*No. of workers*
Below 100	10
Below 150	35
Below 200	180
Below 250	400
Below 300	470
Below 350	500

Calculate Karl Pearson's coefficient of skewness and interpret its value.

46. (i) Mode is more than mean by 4.5 only and the variance is 121, then find out coefficient of skewness.

(ii) Bowley's coefficient of skewness (–) 0.059, Q_1 = 58.24 and Median = 61.8. Find out Q_3.

(iii) Karl Pearson's coefficient of skewness of a distribution is + 0.4, its standard deviations is 6.5 and mean 29.6. Find the Mode and Median of the distribution.

(iv) In a certain distribution, the following results were obtained, $\overline{X} = 90$, mode = 96, Coefficient of Sk. = – 0.4. Find the Standard deviation.

47. Calculate coefficient of quartile deviation and Bowley's coefficient of skewness from the following data :

Profits (Rs. Lakhs) :	Below 10	10–20	20–30	30–40	40–50	Above 50
No. of Cos. :	8	12	20	16	5	2

[Coeff. of Q.D. = 0.313, Coeff. of Sk. = 0.002]

48. From the following information given to you, can it be concluded that the distribution platykurtic :

N = 100, Σ fdx = 50, Σ fdx^2 = 167.2,

Σ fdx^3 = 2925.8, Σ fdx^4 = 86650.2

[β_2 = 2.22, yes]

49. Calculate coefficient of skewness using the mean, mode and standard deviation from the following data :

Marks less than :	20	30	40	50	60	70	80

No. of students : 10 25 40 65 80 85 100

[– 0.289]

50. (a) Calculate the coefficient of skewness with the help of the following data :

Age below :	10	20	30	40	50	60
No. of person :	15	25	40	50	55	60

$$\left[+0.161;\ \text{Use } \frac{3(\overline{X} - \text{Med.})}{\sigma}\right]$$

(b) Calculate Karl Pearson's of skewness for the following distribution :

Monthly Salary (in Rs.)	*No. of salesmen*
400 but less than 600	4
600 but less than 800	10
800 but less than 1,000	19
1,000 but less than 1,200	12
1,200 but less than 1,400	4
1,400 but less than 1,600	1

51. (a) Find four central moments for the following distribution :

X :	1	2	3	4	5	6	7	8	9
f :	1	2	3	4	5	4	3	2	1

(b) For a distribution, the first hour moments about zero are 1, 7, 38 and 155 respectively. Compute the moment coefficients of skewness and kurtosis, (ii) Is the distribution mesokurtic?

52. (a) Calculate Coefficient of Skewness if Mode = 11, Mean = 8, the difference of the two quartiles = 8 and the sum of the two quartiles = 22.

(b) The first four moments of a distribution about the value 4 are – 1.5, 17, – 30 and 108. Calculate the central moments.

53. The first four moments of a distribution about the value 5 of the variables are 2, 20, 40 and 50. Calculate the mean variance, β_1 and β_2 and comment upon the nature of distribution.

$$[\overline{X} = 7,\ \mu_2 = 15,\ \beta_1 = 0.25,\ \beta_2 = 0.64]$$

54. (a) Calculate the Pearson's coefficient of skewness and quartile coefficient of skewness for the following data :

Mid Point X	:	5	10	15	20	25	30	35
Frequency	:	2	8	22	34	16	10	8

[0.386]

55. Calculate the first four moments about the mean for following data and examine for the nature of the distribution :

X :	1	2	3	4	5	6	7	8	9
f :	2	6	13	25	30	22	9	5	2

[$b_1 = 0.03$; $b_2 = 3.00$]

56. From the data given below calculate Karl Pearson's Coefficient of Skewness :

Age	*No. of Persons*	*Age*	*No. of Persons*
20–25	50	40–45	150
25–30	70	45–55	120
30–35	80	50–55	70
35–40	180	55–60	50

[0.156]

57. (a) What is skewness? Explain the main types of showed curves.
(b) Distinguish between skewness and kurtosis.
(c) What is skewness? How does it differ from 'dispersion'? Describe the various measures of skewness.

58. (a) How does 'skewness' differ from 'despression'? Explain the relative measures of skewness.
(b) Define 'Moments'. Establish the relation between the moments about the mean in terms of moments about any arbitrary point.

59. "Measures of centra tendency, dispersion and skewness are complementary to each other n understanding the characteristics of a frequency distribution". Explain.

60. Calculate Bowley's coefficient of skewness from the following data:

Mid-point :	1	2	3	4	5	6	7	8	9	10
Frequency :	2	9	11	14	20	24	20	16	5	2

61. What do you understand by skewness and kurtosis? Point out their role in analysing frequency distribution.

62. (a) Compute Quartile Deviation and coefficient of skewness, given the following values :

Median = 18.8, Q_1 = 14.6, Q_2 = 25.2

[Q.D. = 53; Coeff. of Sk. = 0.208]

(b) Find Quartile coefficient of dispersion if coefficient of skewness = – 36. Median = 16.5, Q_1 = 13.8.

63. The first four moments of a distribution about the value 4 are – 1.5, 17, – 30 and 108. Calculate central moments and skewness. (B.A., B.Sc., Part II, Garhwal Univ., M. Com., Jaipur 1992)

[$\mu_1 = 0$, $\mu_2 = 14.75$, $\mu_3 = 3975$, $\mu_4 = 142.31$].

63. (a) Define skewness. How does it differ from dispersion?

(b) Define the terms skewness. Explain different measures of skewness.

(c) Distinguish between skewness and kurtosis.

(d) Explain the concept of skewness. Draw the sketch of a skewed frequency distribution and show the position of the mean, median and mode when the distribution is asymmetric.

65. Prove that the moment coefficient of kurtosis is not affected by the change of scale.

66. Explain what is meant by skewness. How does it differ from dispersion?

67. (a) Find the four moments about mean from the following data. Also make Sheppard's corrections, if necessary and decide whether it is a platykurtic distribution :

Central size of item :	1	2	3	4	5
Frequency :	2	3	5	4	1

(b) (i) Find the coefficient of skewness from the following information:

Diff. of the quartiles = 8 Mode = 11

Sum of two quartiles = 22 Mean = 8

(ii) Calculate the first four moments about the mean from the following data. Also calculate the value of β_1 and β_2.

Marks :	0–10	10–20	20–30	30–40	40–50	50–60	60–70
No. of students ;	5	12	18	40	15	7	3

[(i) β_1 = 0.02, (ii) β_2 = 3)

(iii) Pearson's coefficient of skewness for a distribution is 0.4 and coefficient of variation is 30%. Its mode is 88. Find the mean and the median.

[Mean = 100, Med. = 96]

68. The following are the data regarding the monthly exports of some companies. Find Pearson's coefficient of skewness.

Monthly exports in Lakh Rs. :	5–10	10–15	15–20	20–25	25–30
No. of companies :	12	18	26	14	6

69. Find the coefficient of skewness and coefficient of kurtosis from the following distribution :

Marks :	20–30	30–40	40–50	50–60	60–70
No. of students :	14	25	36	11	14

70. Find Bowley's Coefficient of Skewness from the following data :

Wages (Rs.)	*f*	*Wages (Rs.)*	*f*
110–115	4	135–140	90
115–120	10	140–145	52
120–125	26	145–150	33
125–130	49	150–155	17
130–135	72	155–160	7

71. Calculate Pearson's coefficient of skewness for the data given below and interpret the calculated value :

Marks :	45–50	51–56	57–62	63–68	69–74
Frequency :	12	17	22	18	11

72. (a) The first four central moments are 0, 2.5, 0.7 and 18.75 respectively. Calculate the measures of skewness and kurtosis, and comment on the nature of the distribution.

(b) The followings figures related to wage payments in two firms A and B :

	Firm A	*Firm B*
Mean	75	80
Median	72	70
Mode	67	52
Quartiles	13	17
Standard Deviation	62 and 78	65 and 85

Compare the features of the two distributions.

73. Compute mean, mode and standard deviation from the frequency distribution given below. Use your results to comment on the skewness or the distribution :

Sales (Rs.)	*No. of Firms*	*Sales (Rs.)*	*No. of Firms*
10000–15000	2	35000–40000	8
15000–20000	8	40000–45000	4
20000–25000	6	45000–50000	3
25000–30000	12	50000–55000	1
30000–35000	7	55000–60000	1

74. The first three moments of a distribution about the value 67 of the variable are 0.45, 8.73, and 8.91. Calculate the second and third central moments and comment upon the nature of distribution.

75. (a) Examine whether the following results of a place of computation for obtaining the second (central) moment are consistent or not.

$N = 100, \Sigma fd = -20, \Sigma fd^2 = 220, \Sigma fd^3 = -50, \Sigma fd^4 = 1.240.$

(b) The first four central moments of distribution are 2, 6, 12 and 100. Find β_1 and β_2.

[(a) No : since σ cannot be negative, (b) ($\beta_1 = 0.667$, $\beta_2 = 2.78$)]

76. Calculate the quartile coefficient of skewness of the following data:

Income (Rs.)	*f*	*Income (Rs.)*	*f*
.501–600	5	901–1,000	208
601–700	17	1,001–1,100	134
701–800	80	1,101–1,200	68
801–900	186	1,201–1,300	418

77. What are skewness and kurtosis? Give some suitable measures for skewness and kurtosis.

78. The first two moments of a distribution about the value 4 of a variable are – 1.5 and .27. It is also known that the median of the distribution is 2.1. Calculate Karl Pearson's coefficient of skewness.

79. Define kurtosis. If $\beta_1 = +1$ and $\beta_2 = 4$ and variance = 9. Find the values of μ_3 and μ_4 and comment upon the nature of the distribution.

80. Define moments. How are the skewness and.kurtosis calculated from moments about the mean? Illustrate your answer with an example.

81. The following table gives the length of life (in hours) of 400 T.V. picture tubes :

Length of Life (in hours)	*Number of Picture tubes*	*Length of Life (in hours)*	*Number of Picture tube*
4000–4199	12	5000–5199	55
4200–4399	30	5200–5399	36
4400–4599	65	5400–5599	25
4600–4799	78	5600–5799	9
4800–4999	90		

Compute mean, standard deviation and coefficient of skewness. Comment on the values obtained.

[$\overline{X}$ = 4845.5, σ = 365.9, Coeff. of Sk. = – 0.014]

3

Association of Attributes

INTRODUCTION

In statistics quantitative character arise in two ways.

1. We may measure the actual magnitude or size of some phenomenon. For example, we may measure their weight, etc., the amount of rainfall in a year. The characteristic of this type of phenomena is that they can be quantitatively measured. Data regarding such phenomena are known as statistics of *variables*. The various statistical techniques like measures of central tendency, dispersion, correlation which have been already discussed.
2. There are certain phenomena like blindness, deafness, etc., which are not capable of direct quantitative measurement. In such cases the quantitative character arises only indirectly in the process of counting. For example, we can determine out of 1,000 persons, how many are blind and how many are not blind but we cannot precisely measure blindness. Such phenomena, where direct quantitative measurement is not possible, *i.e.*, where we can study only the presence or absence of a particular characteristic, are called statistics of attributes.

DIFFERENCE BETWEEN CORRELATION AND ASSOCIATION

The correlation is used to measure the degree of relationship between two such phenomena as are capable of direct quantitative measurement. On the other hand, the method of association of attributes is employed to measure the degree of relationship between two phenomena whose size we cannot measure and where we cannot only determine the presence or absence of a particular attribute.

While dealing with statistics of attributes we have to classify the data. The classification is done on the basis of presence or absence of particular distribute or characteristic. When we are studying only one attribute, two classes are formed—one possessing that attribute and another not possessing.

For example, when we are studying the attribute employment, two classes shall be formed; those who are employed and those who are not employed. When two attributes are studied, for classes shall be formed. If, besides employment, we study the sex-wise distribution, four classes shall be formed; number of males employed, number of females employed, number of males unemployed and number of females unemployed.

It should be noted that in some cases while classifying the attributes no clear-cut definition of an attribute and line of demarcation between classes can be drawn. For example, when the attribute 'employment' is being studied the data are classified into 'Employed' and 'Unemployed'. But there can be further category of those people who are partially employed (*i.e.*, part-time). Also there may be some persons who are employed before the survey but on the date of survey they are unemployed. So we cannot treat them as employed and also as unemployed because there is some difference between those persons who have not got any job, and those who have got some job but were retrenched after some time. Hence, it is absolutely essential to lay down clear-cut definition of the various attributes under study. This is often a difficult task. Hence, this limitation must be kept in mind while studying association between attributes.

NOTATIONS

It is customary to use capital letters A and B to represent the presence of the attributes and the Greek letter 'α' (alpha) and 'β' (beta) to represent absence of the attributes. Thus 'α' = not A and 'β' = not B. For example, if A represents males than 'α' would represent females. Similarly, if B represents literates then β would denote illiterates. The combination of the different attributes is denoted by (AB), (Aβ), (αB) and ($\alpha\beta$). Thus in this example. (AB) would mean number of lterate males and ($\alpha\beta$) illiterate females. The number of observations in different classes is called 'class frequency'. Thus, if the number of illiterate males is 50, the frequency of class (AB) is 50. Class frequencies are denoted by enclosing class notation in brackets like (AB), ($\alpha\beta$), etc. Thus

(A) denotes number of individuals possessing attribute A.

(AB) denotes the number of individuals possessing attributes A and B.

($\alpha\beta$) denotes the number of individuals possessing attributes α and β.

Any letter or combination of letters like A, AB, $\alpha\beta$, etc., by means of which we specify the characters of the members of a class, may be termed a class symbol.

Class Frequencies

The number of observations assigned to any class is termed for the sake of brevity the frequency of the class or the 'class frequency'. Class frequencies are denoted by enclosing the corresponding class symbols in brackets. Thus (B) denotes the number of B's *i.e.*, objects possessing attribute B. (Aβ) the number of Aβ's, *i.e.*, objects possessing attribute A but not B, and so on for any number of attributes.

Order of Classes and Class Frequencies

The order of a class depends upon the number of attributes specified. A class having one attribute is known as the class of the first order, a class having two attributes as class of the second order, and so on. The total number of observations denoted by the symbol N is called the frequency of the zero order since no attributes are specified. Thus we have :

$$N \qquad \text{frequency of the zero order}$$

$$\left.\begin{matrix}(A) & (B)\\ (\alpha) & (\beta)\end{matrix}\right\} \text{frequencies of the first order}$$

$$\left.\begin{matrix}(AB) & (\alpha B)\\ (A\beta) & (\alpha\beta)\end{matrix}\right\} \text{frequencies of the second order}$$

Number of Frequencies

In a study of n attributes the total number of class frequencies is given by 3^n.

(i) For one attribute, the frequencies are $3^1 = 3$.

(ii) For two attributes, total frequencies are $3^2 = 9$. They are in the order $1 + 4 + 4 = 9$.

Any class frequency can always be expressed in terms of class frequencies of higher order since the total number of observations must be equal to the number of A's added to the number of α's *i.e.*

$$N = (A) = (\alpha)$$

Similarly, the number of A's is equal to the number of A's which are B's added to the number of A's which are β's *i.e.*,

$$(A) = (AB) + (A\beta)$$

Similarly, we have $(\alpha) = (\alpha B) + (\alpha\beta)$.

Ultimate Class Frequencies

It is clear from above that every class frequency can be expressed in terms of the frequencies of the highest order, *i.e.* of order n. Any frequency can be analysed into highest frequencies, and the process need stop only then we have reached the frequencies of the highest order. For example, with two attributes

$$\left.\begin{array}{l}(A) = (AB) + (A\beta) \\ (\alpha) = (\alpha B) + (\alpha\beta)\end{array}\right\} \text{ultimate class frequencies.}$$

The classes specified by n attributes, *i.e.*, those of the highest order, are termed the ultimate class-frequencies. A given data can be completely specified if only the ultimate class frequencies are given.

The total number of classes of ultimate order is determined by the formula 2^n when n stands for the number of attributes studied.

The frequencies of the positive, negative and ultimate classes can be known from the following table which is known as the nine square table (since nine squares are formed).

	A	α	Total
B	(AB)	(αB)	(B)
β	(Aβ)	(αβ)	(β)
Total	(A)	(α)	N

From this table certain relationships can be described :

$$(A) = (AB) + (A\beta) \; ; \; (\alpha) = (\alpha B) + (\alpha\beta)$$

$$(B) = (AB) + (\alpha\beta) \; ; \; (\beta) = (A\beta) + (\alpha\beta)$$

$$N = (A) + (a) \text{ Or } N = (B) + (\beta)$$

Or

$$N = (AB) + (A\beta) + (\alpha B) + (\alpha\beta)$$

From these relationships if we know any of the ultimate class frequencies and any other three values, we can find out the frequencies of the remaining classes.

Example 1:

From the following two cases find out whether the data are consistent or not

Cases I. (A) = 100, (B) = 150, (AB) = 60, N = 500

Case II. (A) = 100, (B) = 150, (AB) = 140, N = 500

Solution: Case I

We are given (A) = 100, (B) = 150, (AB) = 60, N = 500

Substituting the values in the nine-square table :

	A	α	Total
B	(AB) 60	(αB) 90	(B) 150
β	(Aβ) 40	(αβ) 310	(β) 350
Total	(A) 100	(α) 400	N 500

From the table

(Aβ) = (A) – (AB) = 100 – 60 = 40

(αβ) = (B) – (AB) = 150 – 60 = 90

(αβ) = (α) – (αB) = 400 – 90 = 310

Since all the ultimate class frequencies are positive we conclude that the given data are consistent.

Case II. Given values are :

(A) = 100, (B) = 150, (AB) = 140, (N) = 500

By putting these values in the nine-square table we can determine the missing value :

	A	α	Total
B	(AB) 140	(αB) 10	(B) 150
β	(Aβ) – 40	(αβ) 390	(β) 350
Total	(A) 100	(α) 400	N 500

From the table

(Aβ) = (A) – (AB) = 100 – 140 = – 40

(αβ) = (B) – (AB) = 150 – 140 = 10

(αβ) = (α) – (αB) = 400 – 10 = 390

Thus one of the ultimate class frequencies, *i.e.*, (Aβ) is negative and hence the given data are inconsistent.

Example 2:

From the following data find out the missing frequencies :

(AB) = 100, (A) = 300, (N) = 1,000, (B) = 600

Solution:

Putting these values in the nine-square table :

	A	α	Total
B	(AB) 100	(αB) 500	(B) 600
β	(Aβ) 200	(αβ) 200	(β) 400
Total	(A) 300	(α) 700	N 1,000

The missing frequencies are (Aβ), (αB). (αβ), (α) and (β),

$$(A\beta) = (A) - (AB) = 300 - 100 = 200$$

$$(\alpha) = N - (A) = 1000 - 300 = 700$$

$$(\beta) = N - (B) = 1000 - 600 = 400$$

$$(\alpha\beta) = (B) - (AB) = 600 - 100 = 500$$

$$(\alpha\beta) = (\beta) - (A\beta) = 400 - 200 = 200$$

Thus the missing frequencies are :

$$(A\beta) = 200, (\alpha\beta) = 500, (\alpha\beta) = 200, (\beta) = 400, (\alpha) = 700$$

CONSISTENCY OF DATA

To find out whether the given data are consistent have to apply a very simple test. The test is to find out whether any one or more of the ultimate class-frequencies is negative or not. If none of the class-frequencies is negative we can safely calculate that the given data are consistent (*i.e.*, the frequencies do not conflict in any way with each other). On the other hand, if any of the ultimate class frequencies comes out to be negative the given data are inconsistent. Thus the necessary and sufficient condition for the consistency of a set of independent class frequencies is that no ultimate class frequency is negative.

ASSOCIATION AND DISASSOCIATION

The word association as used in Statistics has a technical meaning different from the one in ordinary speech. In common language one speaks of A and B as being 'associated' if they appears together in a number of cases. But in Statistics, A and B are associated only if they are independent. On the other hand, if this number (or proportion) is less than expected for independence, they are disassociated. Thus, in case of second order frequencies, A and B are :

(i) associated if $(AB)(\alpha\beta) > (A\beta)(\alpha B)$; and

(ii) associated if $(AB)(\alpha\beta) < (A\beta)(\alpha B)$.

Hence, it should carefully be noted that association cannot be inferred from the mere fact that some A's and B's, however great the proportion.

METHODS OF STUDYING ASSOCIATION

The following methods to ascertain whether two attributes or not :

I. Comparison of Observed and Expected Frequencies Methods
II. Proportion Method
III. Yule's Coefficient of Association
IV. Coefficient of Colligation
V. Coefficient of Contingency.

I. **Comparison of Observed and Expected Frequencies Methods :** When this method, the actual observation is compared with the expectation. If the actual observation is equal to the expectation, the attributes are said to be independent: if actual observation is more than the expectation, the attributes are said to be positively associated and if the actual observation is less than the expectation, the attributes are said to be negatively associated.

Symbolically, attributes A and B are :

(i) Independent if $(AB) = \dfrac{(A) \times (B)}{N}$ (expectation) :
(actual observation)

(ii) positively associated if $(AB) > \dfrac{(A) \times (B)}{N}$ (expectation); and
(actual observation)

(iii) negatively associated if $(AB) < \dfrac{(A) \times (B)}{N}$ (expectation).
(actual observation)

The same is true for attributes α and B ; α and β and A and B. Thus, attributes α and β shall be called :

(i) independent, if $(\alpha\beta) = \dfrac{(\alpha) \times (\beta)}{N}$;

(ii) positively associated, if $(\alpha\beta) > \dfrac{(\alpha) \times (\beta)}{N}$; and

(iii) negatively associated, if $(\alpha\beta) > \dfrac{(\alpha) \times (\beta)}{N}$.

Example 1:

According to a survey the following results were obtained :

	Boys	*Girls*
No. of candidates appeared at an examination	*800*	*200*
Married	*150*	*50*
Married and successful	*70*	*20*
Unmarried and successful	*550*	*110*

Find the association between marital status and the success in the examination both for boys and girls.

Solution:

Let A denote married

$\therefore \alpha$ will denote unmarried

Let B denote success

$\therefore \beta$ will denote not successful

We will find coefficient of association both for boys and girls.

Boys : (A) = 150, (AB) = 70, (αB) = 550, N = 800.

By putting the given values in the nine-square table, we will find out the missing ultimate class frequencies.

	A	α	Total
B	(AB) 70	(αB) 550	(B) 620
β	(Aβ) 80	($\alpha\beta$) 100	(β) 180
Total	(A) 150	(α) 650	N 800

From the table

$$(A\beta) = 80 \text{ and } (\alpha\beta) = 100$$

$$Q = \frac{(AB)(\alpha\beta) - (A\beta)(\alpha B)}{(AB)(\alpha\beta) + (A\beta)(\alpha B)}$$

$$= \frac{(70 \times 100) - (80 \times 550)}{(70 \times 100) + (80 \times 550)}$$

$$= \frac{7000 - 44000}{7000 + 44000} = -0.725$$

Girls. (A) = 50, (AB) = 20, (αB) = 110, N = 200

	A	α	Total
B	(AB) 20	(αB) 110	(B) 130
β	(Aβ) 30	($\alpha\beta$) 40	(β) 70
Total	(A) 50	(α) 150	N 200

$$(A\beta) = 30 \text{ and } (\alpha\beta) = 40$$

$$Q = \frac{(20 \times 40) - (30 \times 110)}{(20 \times 40) + (30 \times 110)}$$

$$= \frac{800 - 3300}{800 + 3300} = \frac{-2500}{4100} = -0.61$$

It is clear from the answer that the marital status and the success in the examination are negatively associated for boys as well as girls.

Example 2:

Survey was conducted in respect of marital status and success in examination. Let of 2000 persons who appeared for an examination, 80% of them were boys, and the were girls. Among 300 married boys, 140 were successful, 1,100 boys were successful among unmarried boys. In respect of 10 married girls 40 were successful, 200 unmarried girls were successful. Construct two separate nine-square tables and find out the Yule's Coefficient of Association to discuss the association between marital status and passing of examination.

Solution:

Let A denote married boys

$\therefore$ α would denote unmarried boys

Let B denote those who were successful

$\therefore$ β will denote those who were unsuccessful

In respect of boys, we are given the following information :

N = 1,600, (A) = 300, (AB) = 140, ($\alpha\beta$) = 1,100

We can find out the missing values from the nine-square table.

	A	α	Total
B	140	1100	1240
β	160	200	360
Total	300	1300	1600

$$\text{Expectation of (AB)} = \frac{(A) \times (B)}{N}$$

(A) = 300, (B) = 1,240, N = 1,600

$$\text{Expectation of (AB)} = \frac{300 \times 1240}{1600} = 232.5$$

Since (AB) actual observation (140) is less than the expectation (232.5) the attributes marriage and success in the examination are negatively associated.

We can construct another table in respect of girls. Taking A as married girls given information is :

N = 400, (AB) = 40, (A) = 100, (αβ) = 200

We can find out the missing frequencies from the nine-square table,

	A	α	Total
B	40	200	240
β	60	100	160
Total	100	300	400

$$\text{Expectation of (AB)} = \frac{(A) \times (B)}{N}$$

$$= \frac{100 \times 240}{400} = 60$$

Since (AB) actual observation (40) is less than the expectation (60) the attributes marriage and success in the examination are negatively associated.

Example 3:

Find whether A and B are independent in the following case :

(AB) = 256, (αB) = 768, (Aβ) = 48, (αβ) = 144.

Solution:

Attributes A and B shall be independent if

$$\frac{(AB)}{(A)} = \frac{(\alpha B)}{(\alpha)}$$

For finding (A), (αB) let us prepare a nine-square table :

	A	α	Total
B	256	768	1024
β	48	144	192
Total	304	912	1216

(AB) = 256, (A) = 304, (αβ) = 768, (α) = 912

$$= \frac{256}{304} = \frac{768}{912}$$

Since left hand and right hand sides are equal :

$$\therefore \quad \frac{(AB)}{(A)} = \frac{(\alpha B)}{(\alpha)}$$

Hence attributes A and B are independent.

Example 4:

In a certain class it was found that 70% of the students passed in half-yearly examination, 30% students passed half-yearly and annual examination, while 28% were such who passed in annual but failed in half-yearly examination. Find the percentage of students who :

(a) passed in annual examination.

(b) passed in half-yearly but failed in annual examination, and

(c) failed in both the examinations.

Solution:

Let A denote those passing annual examination

Let B denote those passing the half-yearly examination, α and β will represent respectively those who fail in the annual examination and those failing in the half-yearly examination.

(i) (A), (ii) (αB), and (iii) (αβ)

(i) (A) = (AB) + (Aβ) = 30 + 28 = 58%

Hence the percentage of students who passed in the annual examination is 58.

(ii) (αB) = (B) – (AB) = 70% – 30% = 40%

Hence the percentage of students who failed in annual but passed in half-yearly is 40.

(iii) $(\alpha\beta) = (\beta) - (A\beta) = N - (B) - (A\beta) = 100 - 70 - 28 = 2.$

Hence 2% students failed in both the examinations.

Example 5:

Do you find any association between the temperaments of brothers and sisters from the following data :

Good natured brothers and good natured sisters	*1,230*
Good natured brothers and sullen sisters	*850*
Sullen brothers and good natured sisters	*530*
Sullen brothers and sullen sisters	*980.*

Solution:

Let A denote good natured brothers, and B denote good natured sisters.

$\therefore$ α will denotes sullen brothers, and β will denote sullen sisters.

We are given (AB) = 1230, $(A\beta) = 850$,

$$(\alpha B) = 530, (\alpha\beta) = 980$$

$$Q = \frac{(AB)(\alpha\beta) - (A\beta)(\alpha B)}{(AB)(\alpha\beta) + (A\beta)(\alpha B)}$$

$$= \frac{(1230)(980) - (850)(530)}{(1230)(980) + (850)(530)}$$

$$= \frac{1205400 - 450500}{1205400 + 450500} = \frac{754900}{1655900} = +0.546.$$

We thus find positive association between the temperaments of brothers and sisters.

Example 6:

As from the following data find out whether attributes (i) (AB), (ii) $(A\beta)$, (iii) (αB) and (iv) $(\alpha\beta)$ are independent, associated or disassociated N = 100, (A) = 40, (B) = 80, (AB) = 30.

Solution:

(i) Apply the criterion of independence, *i.e.*, attributes (AB) shall be called independent if $(AB) = \frac{(A) \times (B)}{N}$;

positively associated if $(AB) > \frac{(A) \times (B)}{N}$; and negatively associated or disassociated if $(AB) < \frac{(A) \times (B)}{N}$.

Expectation of $(AB) = \frac{(A) \times (B)}{N}$. Here (A) = 40, (B) = 80, N = 100.

Expectation of $(AB) = \frac{40 \times 30}{100} = 32$.

The actual observation [*i.e.*, the given value of (AB), *i.e.*, 30] is less than the expectation and hence the attributes are disassociated or negatively associated.

(ii) For finding out the nature of association between the attributes (Aβ), (αβ) and ab we shall have to determine the unknown values. This can be done by preparing a nine-square table :

	A	α	Total
B	30	50	80
β	10	10	20
Total	40	60	100

From the table (Aβ) = 10, (αB) = 50, (αβ) = 10, (α) = 60, (β) = 20.

Attributes A and β shall be independent if $(A\beta) = \frac{(A) \times (\beta)}{N}$

Expectation of $(A\beta) = \frac{(A) \times (\beta)}{N}$, where (A) = 40, (β) = 20, N = 100

$\therefore$ Expectation of $(A\beta) = \frac{40 \times 20}{100} = 8$.

Thus the actual observation [*i.e.*, (Aβ) = 10] is more than expectation and hence the attributes A and β are positively associated.

(iii) Attributes α and B shall be called independent if $(\alpha B) = \frac{(\alpha) \times (B)}{N}$

Expectation of $(\alpha B) = \frac{(\alpha) \times (B)}{N}$, where (α) = 60, (B) = 80, N = 100

$\therefore$ Expectation of $(\alpha B) = \frac{60 \times 80}{100} = 48$.

Thus actual observation [*i.e.*, (αB) = 50] is more than the expectation and hence the attributes are positively associated.

(iv) Attributes α and β shall be called independent if $(\alpha\beta) = \frac{(\alpha) \times (\beta)}{N}$

Expectation of $(\alpha\beta) = \frac{(\alpha) \times (\beta)}{N}$, where (α) = 60, (β) = 20, N = 100

$\therefore$ Expectation of $(\alpha\beta) = \dfrac{60 \times 20}{100} = 12.$

Thus actual observation $(\alpha\beta)$ = 10] is less than the expectation (12). Hence the attributes are disassociated.

Example 7:

Find if A and B are independent, positively associated or negative associated from the data given below :

(A) = 470, (B) = 620, (AB) = 320, N = 1000

Solution:

Attributes A and B shall be called independent it

$$(AB) = \frac{(A) \times (B)}{N}$$

(AB) = 302, (A) = 470, (B) 620, N = 1000

$$\text{Expectation of (AB)} = \frac{470 \times 520}{1000} = 291.4$$

Since (AB) actual observation (320) is more than the expectation (291.4) attributes A and B are positively associated.

Limitations. With the help of this method we can only determine the nature of association (*i.e.*, whether there is positive or negative association or no association) and not the degree of association (*i.e.*, where association is high or low). Yule's coefficient is superior because it provides information not only on the nature but also on the degree of association.

II. **Proportion Method :** If there is no relationship of any kind between two attributes A and B we expect to find the same proportion of A's amongst the B's as amongst the β's. Thus, if a coin is tossed we expect the same proportion of heads irrespective of whether the coin is tossed by the right hand or the left hand.

Symbolically, two attributes may be termed :

(i) independent if $\dfrac{(AB)}{B} = \dfrac{(A\beta)}{(\beta)}$

(ii) positively association if $\dfrac{(AB)}{B} > \dfrac{(A\beta)}{(\beta)}$

(iii) negative associated if $\dfrac{(AB)}{B} < \dfrac{(A\beta)}{(\beta)}$

If the relation (i) holds good the corresponding relations

$$\frac{(\alpha B)}{(B)} = \frac{(\alpha\beta)}{(\beta)};\ \frac{(AB)}{A} = \frac{(\alpha B)}{(\alpha)};\ \frac{(AB)}{(A)} = \frac{(\alpha\beta)}{(\alpha)}$$

must also hold true.

Example 8:

In a population of 500 students the number of married is 200. Out of 150 students who failed 60 belonged to the married group. It is required to find out whether the attributes marriage and failure are independent, positively associated or negatively associated.

Solution:

Let A denote married students.

$\therefore$ α represents unmarried students.

Let B denote number of failures.

$\therefore$ β would denote non-failures.

We are given the total number of students, *i.e.*, N = 500

(A) = 200, (B) = 150 and (AB), *i.e.*, the number of married students who failed = 60.

Applying the proportion method.

Attributes A and B shall be called independent if $\frac{(AB)}{A} = \frac{(\alpha B)}{(\alpha)}$

In other words, if the proportion of married students who failed is the same as the proportion of unmarried students who failed we say that the attributes, marriage and failure, are independent.

Proportion of unmarried students who failed : *i.e.*,

$$\frac{(AB)}{A} = \frac{60}{200} = .3 \text{ or } 30\%$$

Proportion of unmarried students who failed :

i.e., $\frac{(\alpha B)}{(\alpha)} = \frac{90}{300} = 0.3 \text{ or } 30\% \left(\begin{array}{l}(\alpha B) = (B) - (AB) \text{ i.e. } 150 - 60 = 90 \\ (\alpha) = N - (A), \text{ i.e., } 500 - 200 = 300\end{array}\right)$

Since the two proportion are the same we concludes that the attributes, marriage and failure, are independent.

Limitation of the Method. Just like the previous method, in this method also we can only determine the nature of association and not the degree of association.

III. **Yule's Coefficient of Association :** The most popular method of studying association is the Yule's Coefficient because here not only we can determine the nature of association, *i.e.*, whether the attributes are positively associated, negatively associated or independent, but also the degree or extent to which the two attributes are associated. The Yule's Coefficient is denoted by the symbol Q and is obtained by applying the following formula :

$$Q = \frac{(AB)(\alpha\beta) - (A\beta)(\alpha B)}{(AB)(\alpha\beta) + (A\beta)(\alpha B)}$$

The value of this coefficient lies between ± 1. When the value of Q is + 1 there is perfect positive association between the attributes. When Q is – 1 there is perfect negative association (or perfect dissociation) between the attributes and when the value Q is zero the two attributes are independent.

The coefficient of association can be used to compare the intensity of association between two attributes with the intensity of association between two other attributes.

Example 9:

Investigate the association between eye colour of husbands and eye colour of wives from the data given below :

Husband with light eyes and wives with light eyes = *309*

Husband with light eyes and wives with not-light eyes = *214*

Husband with not-light eyes and wives with light eyes = *312*

Husband with not-light eyes and wives with not-light eyes = *119.*

Solution:

Since we have to find out the association between eye colour of husband and that of wife one attribute we would take as A and another as B.

Let A denote husband with light eyes.

∴ α would denote husbands with not-light eyes.

Let B denote wives with light eyes.

∴ β would denote wives with not-light eyes.

The given data in terms of these symbols is

(AB) = 309, (Aβ) = 214, (αB) = 132, (αβ) = 119.

Applying Yule's method : $Q = \frac{(AB)(\alpha\beta) - (A\beta)(\alpha B)}{(AB)(\alpha\beta) + (A\beta)(\alpha B)}$

Thus there is a very little association between the eye colour of husband and wife.

Example 10:

Find the Association between Literacy and Unemployment from the following figures :

Total Adults	*10,000*
Literates	*1,290*
Unemployed	*1,390*
Literate Unemployed	*820*

Comment on the results.

Solution:

Let A denote Literates

$\therefore$ α will denote illiterates

Let B denote Unemployed

$\therefore$ β will denote employed.

We are given : (A) = 1290, (B) = 1390, (AB) = 820, N = 10,000. Putting this information in the nine square tables and finding missing values :

	A	α	Total
B	820	570	1390
β	470	8140	8610
Total	1290	8710	10000

$$Q = \frac{(AB)(\alpha\beta) - (A\beta)(\alpha B)}{(AB)(\alpha\beta) + (A\beta)(\alpha B)}$$

$$= \frac{(820 \times 8140) - (470 \times 570)}{(820 \times 8140) + (470 \times 570)}$$

$$= \frac{6674800 - 267900}{6674800 + 267900} = \frac{6406900}{6942700} = 0.923$$

There is a high degree of positive association between literacy and unemployment.

Example 11:

Eighty-eight residents of an Indian city, who were interviewed during a sample survey, are classified below according to their smoking and tea drinking habits. Calculate Yule's Coefficient of Association and comment on its value.

	Smokers	*Non-Smokers*
Tea Drinkers	*40*	*33*
Non-tea Drinkers	*3*	*12*

Solution:

Let A denote smokers.

∴ α would denote non-smokers.

Let B denote tea drinkers.

∴ β would emote non-tea drinkers.

The given data in terms of these symbols are :

(AB), *i.e.*, number of smokers and tea drinkers = 40

(Aβ), *i.e.*, number of smokers and non-tea drinkers = 3

(αB), *i.e.*, number of tea drinker and non-smokers = 33

(αβ), *i.e.*, number of non-smokers and non-tea drinkers = 12.

Applying Yule's method : $Q = \frac{(AB)(\alpha\beta) - (A\beta)(\alpha B)}{(AB)(\alpha\beta) + (A\beta)(\alpha B)}$

Substituting the values of (AB), (Aβ), (αB) and (αβ) in this formula

$$Q = \frac{(40 \times 12) - (3 \times 33)}{(40 \times 12) + (3 \times 33)}$$

$$= \frac{480 - 99}{480 + 99} = \frac{381}{579} = 0.658.$$

This shows that the attributes tea drinking and smoking are positively associated.

Example 12:

Prepare a 2 × 2 table from the following information, calculate Yule's Coefficient of Association and interpret the result.

N = 1500, (α) = 1117,

(B) 360, (AB) = 35

Solution:

By putting the known values in the nine-square table, we can find out the unknown values.

	A	*α*	*Total*
B	35	325	360
β	348	792	1140
Total	383	1117	1500

Thus $(A) = N\ (\alpha) = 1500 - 1117 = 383$

$(A\beta) = (A) - (AB) = 383 - 35 = 348$ etc.

Yule's Coefficient of Association

$$Q = \frac{(AB)(\alpha\beta) - (A\beta)\ (\alpha B)}{(AB)(\alpha\beta) + (A\beta)\ (\alpha B)}$$

$$= \frac{(35 \times 792) - (348 \times 325)}{(35 \times 792) + (348 \times 325)}$$

$$= \frac{27720 - 113100}{27720 + 113100} = \frac{-85380}{140820} = 0.606.$$

Criteria of Independence

The following is the list of criteria that may be used in order to ascertain whether attributes A and B are independent. A and B would be independent if :

$$\frac{(AB)}{B} = \frac{(A\beta)}{(\beta)} \text{ or } \frac{(\alpha B)}{(B)} = \frac{(\alpha\beta)}{(\beta)}$$

or $$\frac{(AB)}{A} = \frac{(\alpha\beta)}{(\alpha)} \text{ or } \frac{(AB)}{(A)} = \frac{(\alpha\beta)}{(\alpha)}$$

or $$\frac{(AB)}{(B)} = \frac{(A)}{N} \text{ or } \frac{(AB)}{(A)} = \frac{(B)}{N}$$

or $$(AB) = \frac{(A)(B)}{N} \text{ or } \frac{(AB)}{N} = \frac{(A)}{N} \times \frac{(B)}{N}$$

or $$(AB)\ (ab) = (AB)\ (\alpha\beta)$$

or $$\frac{(AB)}{(\alpha\beta)} = \frac{(A\beta)}{(\alpha\beta)}$$

or $$\frac{(AB)}{(A\beta)} = \frac{(\alpha B)}{(\alpha\beta)}$$

IV. **Coefficient of Colligation** Yule has computed another coefficient called the coefficient of 'colligation'. It is denoted by the symbol γ and is obtained by applying the following formula :

$$\gamma = \frac{1 - \sqrt{\dfrac{(A\beta)(\alpha B)}{(AB)(\alpha\beta)}}}{1 + \sqrt{\dfrac{(A\beta) \times (\alpha B)}{(AB) \times (\alpha\beta)}}}$$

From this coefficient we can obtain Yule's Coefficient of Association, *i.e.*, Q as follows :

$$Q = \frac{2\gamma}{1 + \gamma^2}$$

It should be noted that though γ and Q serve the same purpose, these coefficients are not directly comparable with each other. Further, in practice Q is more popularly used than γ as a measure of association.

V. **Coefficient of Contingency** So far we have considered cases of dichotomous classification. However, qualitative data are often classified into more than two classes, *i.e.*, attribute A may be classified not as 'A' and 'not A', but as A_1, A_2, A_3 etc. Similarly, another attribute B may be subdivided into B_1, B_2, B_3 etc. The frequencies falling within the different classes may be arranged in the form of a Contingency Table as follows :

B / **A**	**B_1**	**B_2**	**B_3**	**...**	**B_n**	**Total**
A_1	$(A_1 B_1)$	$(A_1 B_2)$	$(A_1 B_3)$	...	$(A_1 B_n)$	(A_1)
A_2	$(A_2 B_2)$	$(A_2 B_2)$	$(A_2 B_3)$	...	$(A_2 B_n)$	(A_2)
A_3	$(A_3 B_1)$	$(A_3 B_2)$	$(A_3 B_3)$	...	$(A_3 B_n)$	(A_3)
...	...	...	...	...	...	...
A_n	$(A_n B_1)$	$(A_n B_2)$	$(A_n B_3)$	...	$(A_n B_n)$	(A_1)
Total	(B_1)	(B_2)	(B_3)	...	(B_n)	N

For determining the degree of association between A's and B's on the whole the coefficient of mean square contingency as given by Pearson may be used. The coefficient of mean square contingency is denoted by the symbol i and obtained by applying the following formula :

$$C = \sqrt{\frac{\chi^2}{N + \chi^2}}$$

While finding out the value of i we proceed on the assumption of null hypothesis, *i.e.*, the two attributes are independent and exhibit no association.

For calculation of i we have to determine the value of χ^2 (pronounced as chi-square). The step in calculating the value of χ^2.

(i) Find the expected or independent frequency for each cell. Thus, for cell (A_1B_1) the expectation is $\frac{(A_1) \times (B_1)}{N}$.

(ii) Obtain the difference between the observed (actual) and expected frequencies in each cell, *i.e.*, find (O – E).

(iii) Square (O – E) and divide the figure by E, the expected frequency for each cell.

(iv) Add up the figures obtained in step (iii). This would give the value of χ^2. Thus $\chi^2 = \Sigma \frac{(O - E)^2}{E}$.

Once the value of χ^2 is obtained it is easy to determine the value of i.

Example 13:

In two towns A and B, the following information was supplied by an investigator :

		Town A	*Town B*
Total population (in thousands)		*240*	*234*
Literates	*(")*	*40*	*34*
Illiterate criminals	*(")*	*40*	*20*
Literate criminals	*(")*	*5*	*2*

Compare the degree of association between literacy and crime in each of the two towns.

Solution:

Let A denote literates

∴ α will denote illiterates

Let B denote criminals

∴ β will denote non-criminals

In terms of these symbols the given information is expressed as below:

	Town A	*Town B*
N	240	234
(A)	40	34
(αB)	40	20
(AB)	5	2

The missing frequencies can be ascertained by the nine-square table

Town A

	A	α	Total
B	(AB) 5	(αB) 40	(B) 45
β	(Aβ) 35	(αβ) 160	(β) 195
Total	(A) 40	(α) 200	N 240

Town B

	A	α	Total
B	(AB) 2	(αB) 20	(B) 22
β	(Aβ) 32	(αβ) 180	(β) 212
Total	(A) 34	(α) 200	N 234

Town A $\quad Q = \dfrac{(AB)(\alpha\beta) - (A\beta)(\alpha B)}{(AB)(\alpha\beta) + (A\beta)(\alpha B)}$

$$= \frac{(5) \times (160) - (35) \times (40)}{(5) \times (160) - (35) \times (40)}$$

$$= \frac{800 - 1400}{800 + 1400} = \frac{-600}{2200} = 0.273$$

Town B $\quad Q = \dfrac{(2) \times (180) - (32) \times (20)}{(2) \times (180) + (32) \times (20)}$

$$= \frac{360 - 640}{360 + 640} = \frac{-280}{1000} = -0.28.$$

Thus there is dissociation between literacy and criminality in both the towns. The degree of dissociation is slightly more in town B compared to town A.

Example 14:

In a co-educational institution, out of 200 students 150 were boys. They took an examination and it was found that 120 passed, 10 girls had failed. Is there any association between sex and success in the examination?

Solution:

Let A denote boys

∴ α will denote girls.

Let B denote those who passed the examination.

∴ β will denote those who failed.

We are given : N = 200, (A) = 150, (AB) = 120, (αβ) = 10.

Other frequencies can be obtained from the nine-square table.

	A	α	Total
B	120	40	160
β	30	10	40
Total	150	50	200

Applying Yule's Method $Q = \dfrac{(AB)(\alpha\beta) - (A\beta)(\alpha B)}{(AB)(\alpha\beta) + (A\beta)(\alpha B)}$

$$(AB) = 120,\ (\alpha\beta) = 20,\ (A\beta) = 30,\ (\alpha B) = 40$$

$$Q = \frac{(120 \times 10) - (30 \times 40)}{(120 \times 10) + (30 \times 40)}$$

$$= \frac{1200 - 1200}{1200 + 1200} = 0$$

Hence there is no association between sex and success in the examination.

Example 15:

The male population of U.P. is 250 lakhs. The number of literate males is 20 lakhs and the total number of criminals is 26 thousand. The number of literate male criminals is 2 thousand. Do you find any association between literacy and criminally?

Solution:

Let A denote literate males

∴ α will denote illiterate males

Let B denote males criminals

∴ β will denote male non-criminals

The given information in lakhs is

$N = 250,\quad (A) = 20,$

$(B) = 0.26,\ (AB) = 0.02$

$$\text{Expectation of } (AB) = \frac{(A) \times (B)}{N} = \frac{20 \times .26}{250} = .0208$$

Since (AB) actual observation (0.02) is less than expectation (0.0208), the attributes literacy and criminality are negatively associated, *i.e.*, literacy checks criminality.

Example 16:

In a group of 800 students, the number of married is 320. But of 240 students who failed, 96 belonged to the married group. Find out whether the attributes manage and failure are independent.

Solution:

Let A stand for married students and B for those who have failed. We are given N = 800, (A) = 320, (B) = 240, (AB) = 96. By putting the information in nine-square table, we have

	A	α	Total
B	96	144	240
β	224	336	560
Total	320	480	800

Calculating Yule's Coefficient of Association

$$Q = \frac{(AB)(\alpha\beta) - (A\beta)(\alpha B)}{(AB)(\alpha\beta) + (A\beta)(\alpha B)}$$

$$Q = \frac{(96 \times 336) - (224 \times 144)}{(96 \times 336) + (224 \times 144)}$$

$$= \frac{32256 - 32256}{32256 + 32256} = 0$$

Example 17:

In an assorted study to find whether tall husbands tend to marry tall wives, the following information about the wives of 1250 tall and 1250 short statured husband was published. Find the coefficient of association between the stature of wives and husbands

	Tall husband	*Short husbands*
Tall wives	*60%*	*10%*
Short wives	*10%*	*50%*

Solution:

Let A denote tall husbands

∴ α will denote short husbands

Let B denote tall wives

∴ β denote short wives

(AB) = 60, (Aβ) = 10, (αB) = 10, (αβ) = 50.

Applying Yule's Coefficient : $Q = \frac{(AB)(\alpha\beta) - (A\beta)(\alpha B)}{(AB)(\alpha\beta) + (A\beta)(\alpha B)}$

$$= \frac{(60 \times 50) - (10 \times 10)}{(60 \times 50) + (10 \times 10)}$$

$$= \frac{3000 - 100}{3000 + 100} = \frac{2900}{3100} = +0.935.$$

Example 18:

Calculate Yule's Coefficient of Association between marriage and failure of students from the following data pertaining to 525 students :

	Passed	*Failed*	*Total*
Married	*90*	*65*	*155*
Unmarried	*260*	*110*	*370*

Solution:

Let A denote married persons

$\therefore$ α will denote unmarried persons

Let B denote those who failed

$\therefore$ β will denote those passed

Thus $(A\beta) = 90$, $(\alpha\beta) = 260$, $(AB) = 65$, $(\alpha B) = 110$

Applying Yule's Coefficient : $Q = \frac{(AB)(\alpha\beta) - (A\beta)(\alpha B)}{(AB)(\alpha\beta) + (A\beta)(\alpha B)}$

$$= \frac{(65) \times (260) - (90) \times (110)}{(65) \times (260) + (90) \times (110)}$$

$$= \frac{16900 - 9900}{16900 + 9900} = \frac{-7000}{26800} = +0.261.$$

Example 19:

The following table shows the association among 1,000 criminals between their weight and mentality. Calculate the coefficient of contingency between the two.

			Weight in Pounds			
Mentality	***110–120***	***120–130***	***130–140***	***140–150***	***Above 150***	***Total***
Normal	*50*	*102*	*198*	*210*	*240*	*800*
Week	*30*	*38*	*72*	*30*	*30*	*200*
Total	*80*	*140*	*270*	*240*	*270*	*1000*

Solution:

Coefficient of contingency of $C = \sqrt{\dfrac{\chi^2}{N + \chi^2}}$; $\chi^2 = \Sigma \dfrac{(O - E)^2}{E}$

Mentality	**Weight in Pounds → B**					
A	**110–120** B_1	**120–130** B_2	**130–140** B_3	**140–150** B_4	**Above 150** B_5	**Total**
Normal (A_1)	50	102	198	210	240	800
Weak (A_2)	30	38	72	30	30	200
Total	80	140	270	240	270	1,000

The expected frequency corresponding to the cell (A_1 B_1) is

$$\frac{(A_1) \times (B_1)}{N} = \frac{800}{1000} \times 80 = 64.$$

The expected frequency corresponding to the cell (A_1B_2) is

$$\frac{(A_1) \times (B_2)}{N} = \frac{800}{1000} \times 140 = 112.$$

The expected frequency corresponding to cell (A_1, B_3) is

$$\frac{(A_1) \times (B_3)}{N} = \frac{800}{1000} \times 270 = 216.$$

The expected frequency correspond to cell (A_1, B_4) is

$$\frac{(A_1) \times (B_4)}{N} = \frac{800}{1000} \times 240 = 192.$$

The expected frequency correspond to cell (A_1, B_5) is

$$\frac{(A_1) \times (B_5)}{N} = \frac{800}{1000} \times 270 = 216.$$

Thus the table of expected frequencies is :

Mentality	**Weight in Pounds B**					**Total**
A	B_1	B_2	B_3	B_4	B_5	
A_1	64	112	216	192	216	800
A_2	16	28	54	48	54	200
Total	80	140	270	240	270	1,000

	O	E	$(O-E)^2$	$(O-E)^2/E$
$(A_1\ B_1)$	50	64	196	3.062
$(A_1\ B_2)$	102	112	100	0.893
$(A_1\ B_3)$	198	216	324	1.500
$(A_1\ B_4)$	210	192	324	1.688
$(A_1\ B_5)$	240	216	576	2.667
$(A_2\ B_1)$	30	16	196	12.250
$(A_2\ B_2)$	38	28	100	3.571
$(A_2\ B_3)$	72	54	324	6.000
$(A_2\ B_4)$	30	48	324	6.750
$(A_2\ B_5)$	30	54	576	10.667
				$\chi^2 = \Sigma\ [(O-E)^2/E = 49.048]$

$$C = \sqrt{\frac{\chi^2}{N+\chi^2}} = \sqrt{\frac{49.048}{1000 + 49.048}}$$

$$= \sqrt{\frac{49.048}{1049.048}} = \sqrt{0.0468} = 0.216$$

Example 20:

From the following ultimate class frequencies of the positive and negative cases and the total number of observations :

$(AB) = 100$, $(\alpha B) = 60$, $(A\beta) = 50$, $(\alpha\beta) = 40$.

Solution:

Substituting the given values in the nine-square table :

	A	α	Total
B	(AB) 100	(αB) 80	(B) 180
β	(Aβ) 50	(αβ) 40	(β) 90
Total	(A) 150	(α) 120	N 270

Frequenc is of Positive Classes

$(A) = (AB) + (A\beta) = 100 + 50 = 150$

$(B) = (AB) + (\alpha B) = 100 + 80 = 180$

Frequencies of Negative Classes

$(\alpha) = (\alpha B) + (\alpha\beta) = 80 + 40 = 120$

$(\beta) = (A\beta) + (\alpha\beta) = 50 + 40 = 90$

$N = (AB) + (A\beta) + (\alpha B) + (\alpha\beta) = 100 + 50 + 80 + 40 = 270.$

MISCELLANEOUS EXAMPLES

Example 1:

Comment on the following statements:

(a) *"99% of the people who drink beer die before reaching 100 years of age. Therefore, drinking beer is bad for iongevity."*

(b) *"Road accident resulted in 5,012 deaths in 1988 in India and 8,623 in 1989 while the number of women drivers increased in the period. Hence women make bad drivers."*

Solution:

(a) We are not given complete information, *i.e.*, what percentage of people who do not drink beer die before reaching 100 years of age and as such the inference drawn above is wrong. It is possible that 100% of the persons who do not drink beer may die before reaching 100 years of age in which case drinking may be found to be good for longevity. Therefore, for association between A and B in addition to $\frac{(AB)}{(A)}$, it is necessary to know $\frac{(\alpha B)}{(\alpha)}$ also.

(b) On the basis of the information given it cannot be concluded that the women make bad drivers. Women can be regarded as bad drivers only if number of accidents made per woman driver is more than the number of accidents made by a man driver. These figures are not given. The increase in the number of accidents from 5,082 in 1988 to 8,623 in 1989 may be not due to bad driving but due to other factors like increase in population, increase in number of vehicles on the road, more congested roads, etc. Hence the statement is an example of illusory association.

Example 2:

From the following data, find the association between darkness of eye-colour in father and son :

Father with dark eyes and sons with dark eyes	*1040*
Father with dark eyes and sons with not dark eyes	*160*

Father with not dark eyes and sons with dark eyes 180

Father with not dark eyes and sons with not dark eyes 180

What would have been the frequency of fathers with dark eyes and sons with dark eyes for the same total number, had there been complete independence ?

Solution:

Let A denote fathers with dark eyes

$\therefore$ α would denote fathers with not dark eyes

Let Be denote sons with dark eyes

$\therefore$ β would denote sons with not dark eyes.

The given data in terms of these symbols is:

$(AB) = 1040, (A\beta) = 160, (\alpha B) = 180, (\alpha\beta) = 120$

Applying Yule's Coefficient :

$$Q = \frac{(AB)(\alpha\beta) - (A\beta)(\alpha B)}{(AB)(\alpha\beta) + (A\beta)(\alpha B)}$$

$$= \frac{(1040 \times 120) - (160 \times 180)}{(1040 \times 120) + (160 \times 180)}$$

$$= \frac{124800 - 28800}{124800 + 28800} = \frac{96000}{153600} = +0.625.$$

For finding out the frequency of fathers with dark eyes and sons with dark eyes for the same total number had there been complete independence we shall find out expectation of (AB).

$$\text{Expectation of (AB)} = \frac{(A) \times (B)}{N}$$

For calculating (A), (B) and N we will put the given information in nine square table.

	A	α	Total
B	1040	180	1220
β.	160	120	280
Total	1200	300	1500

A = 1200, B = 1220, N = 1500

$$\text{Hence expectation of (AB)} = \frac{1200 \times 1200}{1500} = 976.$$

Example 3:

A report gives the following observed frequencies :

N = 1000, (A) = 510, (B) = 490, (C) = 427,

(AB) = 189, (AC) = 140, (BC) = 185.

Show that there is some misprint or mistake of some sort and possibly the misprint consists in dropping of 1 before 85 given as frequency of (BC).

Solution:

Let (A) denote manured fields. (B) irrigated fields and (C) fields growing improved varieties.

The given data are : (A) = 510, (B) = 490, (C) = 427,

(AB) = 189, (AC) = 140

From the condition of consistency we have :

$$(AB) + (BC) + (AC) \leq (A) + (B) + (C) - N$$

or $(BC) = (A) + (B) + (C) - (AB) - (AC) - N$

$$85 \geq 510 + 490 + 427 - 189 - 140 - 1000 \geq 98.$$

But the given values of (BC) is 85 which is less than 98. Hence 85 cannot be the correct value of (BC).

Perhaps it is 185 because if (BC) is taken equal to 185, the remaining conditions are satisfied.

$$(AB) + (AC) - (BC) = 189 + 140 - 185$$

$$= 144 \text{ which is less than (A), } i.e., 513$$

$$(AB) + (BC) - (AC) = 189 + 185 - 140 = 234$$

which is less than (B), *i.e.*, 490 and

$$(BC) + (AC) - (AB) = 185 + 140 - 189 = 136$$

which is less than (C), *i.e.*, 427.

Example 4:

Calculate coefficient of partial association between A and B and A and C from the following data :

$$(ABC) = 200,\ (AB\gamma) = 210,\ (A\gamma C) = 208,\ (A\beta\gamma) = 190,$$

$$(\alpha BC) = 209,\ (\alpha B\gamma) = 105,\ (\alpha\beta C) = 170,\ (\alpha\beta\gamma) = 178.$$

Solution:

(i) Between A and B for sub-population C

$$Q_{AB.C} = \frac{(ABC)(\alpha\beta C) - (A\beta C)(\alpha BC)}{(ABC)(\alpha\beta C) + (A\beta C)(\alpha BC)}$$

$$= \frac{(200 \times 170) - (208 \times 209)}{(200 \times 170) - (208 \times 209)} = \frac{-9{,}472}{77{,}472} = 0.122$$

(ii) Between A and B for sub-Population γ

$$Q_{AB.\gamma} = \frac{(AB\gamma)(\alpha\beta\gamma) - (A\beta\gamma)(\alpha B\gamma)}{(AB\gamma)(\alpha\beta\gamma) + (A\beta\gamma)(\alpha B\gamma)}$$

$$= \frac{(210 \times 178) - (190 \times 105)}{(210 \times 178) - (190 \times 105)}$$

$$= \frac{37{,}380 - 19{,}950}{37{,}380 + 19{,}950} = \frac{17{,}430}{57{,}330} = 0.304$$

(iii) Between A and C for sub-population B

$$Q_{AC.B} = \frac{(ABC)(\alpha\beta\gamma) - (AB\gamma)(\alpha BC)}{(ABC)(\alpha\beta\gamma) + (AB\gamma)(\alpha BC)}$$

$$= \frac{(200 \times 105) - (210 \times 209)}{(200 \times 105) + (210 \times 209)}$$

$$= \frac{21{,}000 - 43{,}890}{21{,}000 + 43{,}890} = \frac{-22{,}890}{64{,}890} = -0.353$$

(iv) Between A and C for sub-population b

$$Q_{AC.\beta} = \frac{(A\beta C)(\alpha\beta\gamma) - (A\beta\gamma)(\alpha\beta C)}{(A\beta C)(\alpha\beta\gamma) + (A\beta\gamma)(\alpha\beta C)}$$

$$= \frac{(208 \times 178) - (190 \times 170)}{(208 \times 178) + (190 \times 170)}$$

$$= \frac{37{,}024 - 32{,}300}{37{,}024 + 32{,}300} = \frac{4{,}724}{69{,}324} = +0.068.$$

Example 5:

1800 candidates appeared for an examination, 450 were successful, 340 had attended a coaching class and out of these 200 came out successful. Estimate the utility of coaching class (use Yule's Coefficient of Association).

Solution:

The above information can be summarised in the table given below:

	Success	*No Success*	*Table*
Spl. Coaching	200	140	340
No. Coaching	250	1210	1460
Total	240	1350	1800

$$Q = \frac{(AB)(\alpha\beta) - (A\beta)(\alpha B)}{(AB)(\alpha\beta) + (A\beta)(\alpha B)}$$

$$Q = \frac{(200 \times 1210) - (250 \times 140)}{(200 \times 1210) + (250 \times 140)}$$

$$= \frac{242000 - 35000}{242000 + 35000} = \frac{207000}{277600} = +0.747.$$

Example 6:

A survey of 1,000 companies gives the following data :

(i)	*Companies with a capital of more than Rs. 10 lakhs*	*510*
(ii)	*Companies making profits*	*490*
(iii)	*Companies under managing agents*	*427*
(iv)	*Companies with a capital of more than Rs. 10 lakhs and making profits*	*189*
(v)	*Companies with a capital of more than Rs. 10 lakhs and under managing agents*	*140*
(vi)	*Companies making profits and under managing agents*	*85*

Show that the information as it stands must be incorrect.

Solution:

Let A denote companies with a capital of more than Rs. 10 lakhs, B denote companies making profits, and C denote companies under managing agents.

We are given :

$$N = 1000, (A) = 510, (B) = 490, (C) = 427$$

$$(AB) = 189, (AC) = 150, (BC) = 85$$

The condition of consistency when positive class frequencies are given except (ABC) is:

$$(AB) + (AC) + (BC) < (A) + (B) + (C) - N$$

Substituting the given values

189 + 140 + 85 < 510 + 490 + 427 – 1000 = 414 $\nless$ 427 which is not possible.

Hence the information as it stands must be incorrect.

Example 7:

When are two attributes said to be independent?

(b) From the following data calculate the remaining class frequencies

$$N = 100,\ (A) = 50,\ (B) = 70 \text{ and } (AB) = 30.$$

(c) Use the data in (b) to test whether the attributes A and B are independent or not.

Solution:

(b) Calculation of missing frequency based on the given data

	A	α	Total
B	30	40	70
β	20	10	30
Total	50	50	100

$$Q = \frac{(AB)(\alpha\beta) - (A\beta)(\alpha B)}{(AB)(\alpha\beta) + (A\beta)(\alpha B)}$$

$$= \frac{(30 \times 10) - (20 \times 40)}{(30 \times 10) + (20 \times 40)}$$

$$= \frac{3000 - 800}{300 + 800} = \frac{-500}{1100} = -0.455$$

Example 8:

The male population of a certain State in India is 331 lakhs. The number of literate males is 66 lakhs and the number of male criminals is 33 thousand. If the number of literate male criminals is 6 thousand, calculate the co-efficient of association between literacy and criminality in this State.

Solution:

Let A denote literate males and B denote criminals α and β will denote illiterate males and those who are not criminals respectively. The data given are :

$$N = 3{,}31{,}00{,}000;\ (A) = 66{,}00{,}000;\ (B) = 33{,}000;\ (AB) = 6{,}000$$

Putting these value in the nine-square table

Hence there 10 no association between literacy and employment

6,000	27,000	33,000
65,94,000	2,64,73,000	3,30,67,000
66,00,000	2,65,00,000	3,31,00,000

$$Q = \frac{(AB)(\alpha\beta) - (A\beta)(\alpha B)}{(AB)(\alpha\beta) + (A\beta)(\alpha B)}$$

$$= \frac{(6{,}000)\ (2{,}64{,}73{,}000) - (65{,}94{,}000)\ (27{,}000)}{(6{,}000)\ (2{,}64{,}73{,}000) + (65{,}94{,}000)\ (27{,}000)}$$

$$= \frac{-17{,}20{,}00{,}00{,}000}{3{,}34.87{,}60{,}00{,}000} = -0.056.$$

Thus, there is a low degree of negative association between literacy and criminality.

Example 9:

Prepare a 2 × 2 table from the following information and calculate Yule's Coefficient of Association and interpret the result :

N = 1500, (A) 1117, (B) = 360, (AB) = 35

Solution:

2 × 2 table

	A	α	Total
B	35	325	360
β	348	792	1140
Total	383	1117	1500

$$Q = \frac{(AB)(\alpha\beta) - (A\beta)(\alpha B)}{(AB)(\alpha\beta) + (A\beta)(\alpha B)}$$

$$= \frac{(35)\ (792) - (325)\ (348)}{(35)\ (792) + (325)\ (348)}$$

$$= \frac{27720 - 113100}{27720 + 113100} = \frac{85380}{140820} = -0.606.$$

There is a negative association between the attributes A and B.

Example 10:

Complete the following table and find Yule's coefficient of association.

N = 800; (A) = 470, (β) = 450 and (AB) = 230

Solution:

Putting the given information in a nine-square table and calculating Yule's coefficient of association:

	A	α	Total
B	230	120	350
β	240	210	450
Total	470	330	800

$$Q = \frac{(AB)(\alpha\beta) - (A\beta)(\alpha B)}{(AB)(\alpha\beta) + (A\beta)(\alpha B)}$$

$$= \frac{(230 \times 210) - (240 \times 120)}{(230 \times 210) + (240 \times 120)}$$

$$= \frac{48300 - 28800}{48300 + 28800} = \frac{19500}{77100} = 0.253$$

Example 11:

Calculate the coefficient of association between intelligence in fathers and sons from the following data :

Intelligent fathers with intelligent sons	=	*284*
Intelligent fathers with dull sons	=	*81*
Dull fathers with intelligent sons	=	*92*
Dull fathers with dull sons	=	*579.*

Solution:

Let intelligent fathers be denoted by A and intelligent sons by B. a and b will denote dull father and dull sons respectively. In terms of these symbols, we have

(AB) = 248, (ab) = 81, (aB) = 92, (ab) = 579

Applying Yule's method

$$Q = \frac{(AB)(\alpha\beta) - (A\beta)(\alpha B)}{(AB)(\alpha\beta) + (A\beta)(\alpha B)}.$$

$$= \frac{(248 \times 579) - (81 \times 92)}{(248 \times 579) + (81 \times 92)}$$

$$= \frac{143.592 - 7{,}452}{143{,}592 + 7{,}452} = \frac{136{,}140}{151{,}044} = -0.901.$$

Example 12:

300 people of German and French nationalities were interviewed for finding their preference for music of their own language. The following facts were gathered out of 100 German nationals, 60 liked music of their own language, whereas 70 French nationals out of 200 liked German music. Out of 100 French nationals, 55 liked music of their own language and 35 German nationals out of 200 Germans liked French music.

Using coefficient of association, state whether Germans prefer their own music in comparison with Fenchmen.

Solution:

We prepare the following nine square table for calculating the preference of Germans.

	German Music	*Non-German Music*	*Total*
German Nationals	60	40	100
Non-German Nationals	70	130	200
Total	130	170	300

Yule's coefficient of association between German music and German nationals is

$$Q = \frac{(60 \times 130) - (70 \times 40)}{(60 \times 130) - (70 \times 40)} = \frac{5{,}000}{10{,}600} = +\,0.472$$

Nine square table for calculating the preference of frenchmen is obtained as follows :

	French Music	*Non-French Music*	*Total*
French Nationals	55	45	100
Non-French Nationals	35	165	200
Total	90	210	300

$$Q = \frac{(55 \times 165) - (45 \times 35)}{(55 \times 165) + (45 \times 35)} = \frac{7{,}500}{10{,}650} = 0.704$$

Hence Frenchmen prefer their own music in comparison with Germans.

Example 13:

From the following ultimate class frequencies find the frequencies of the positive and negative classes and the total number of observations.

(AB) = 100, (aB) = 80, (Ab) = 50, (ab) = 40.

Solution:

By putting these values in the nine square table, we can find the desired information.

	A	α	Total
B	100	80	180
β	50	40	90
Total	150	120	270

$N = (AB) + (A\beta) + (\alpha B) + (\alpha\beta)$

$= 100 + 50 + 80 + 40 = 270$

$(A) = (AB) + (A\beta) = 100 + 50 = 150$

$(B) = (AB) + (\alpha B) = 100 + 80 = 180$

$(\alpha) = (\alpha B) + (\alpha\beta) = 80 + 40 = 120$

$(\beta) = (A\beta) + (\alpha\beta) = 50 + 40 = 90$

Example 14:

The following summary data relate to the adult population of a small village.

Adult population	*600*
Number of employed	*240*
Literate adult population employed	*80*
Number of literates	*200*

Determine whether literacy and employment are associated or not.

Solution:

Let A denote literates

∴ α will denote illiterates

Let B denote those who are employed

∴ β will denote those who are unemployed

In terms of these symbols, we have (AB) = 80, (B) = 240, (A) = 200, N = 600. By putting this information in a nine square table, we can find out the missing values.

	A	α	Total
B	80	160	240
β	120	240	360
Total	200	400	600

Applying Yule's method

$$Q = \frac{(AB)(\alpha\beta) - (A\beta)(\alpha B)}{(AB)(\alpha\beta) + (A\beta)(\alpha B)}$$

$$= \frac{(80 \times 240) - (120 \times 160)}{(80 \times 240) + (120 \times 160)}$$

$$= \frac{19200 - 19200}{19200 + 19200} = \frac{0}{38400} = 0$$

Example 15:

With a view to study whether the working condition in a factory had any influence on he frequency of accidents, a researcher collected and tabulated the accident data as follows :

Working Condition	*No. of Accidents*		*Total*
	Less	*More*	
Good	*280*	*80*	*360*
Bad	*120*	*120*	*240*
Total	*400*	*200*	*600*

Using Yule's methodology, calculate the coefficient of association between the number of accidents and the working condition in the factory. What inference would you draw from the result?

Solution:

We are given (AB) = 280, (Aβ) = 120, (αB) = 80, (αβ) = 120

$$Q = \frac{(AB)(\alpha\beta) - (A\beta)(\alpha B)}{(AB)(\alpha\beta) + (A\beta)(\alpha B)}$$

$$= \frac{(280 \times 120) - (120 \times 80)}{(280 \times 120) + (120 \times 80)}$$

$$= \frac{32600 - 9600}{32600 + 9600} = \frac{24000}{43200} = 0.556.$$

There is a moderate degree of positive association between working condition and accidents, *i.e.*, good working condition should generally led to less accident.

Example 16:

Test the consistency of the data given below :

Case = I = (AB) = 200, (A) = 300, (α) = 200, (B) = 250, (αβ) = 150, (N) = 500

Case II = (AB) = 250; (A) = 150, (αB) = 1000, (αβ) = 600, (β) = 500, (N) = 1750.

Solution:

Case I : (AB) = 200, (A) = 300, (α) = 200,

(B) = 250, (αβ) = 150, N = 500.

Putting the given information in a nine square table

200	50	250
100	150	250
300	200	500

Since all the ultimata class frequencies are positive there is n inconsistency in the data

Case II : (AB) = 250, (A) = 150, (αB) = 1000, (αβ) = 600,

(β) = 500, N = 1750

Putting the given information in a nine square table :

250	1000	1250
–100	600	500
150	1600	1750

Since one of the ultimate class frequencies is negative there is some inconsistency in the given information.

Example 17:

Explain from the data given below whether A and B are independent, positively associated or negatively associated :

(i) N = 10,000, (A) = 500, (B) = 6000, (AB) = 3150.

(ii) (A) = 150, (α) = 250, (B) = 200, (αB) = 125

Solution:

(i) Attributes A and B shall be independent if :

$\frac{(AB)}{(A)} = \frac{(\alpha B)}{(\alpha)}$, positively associated, if $\frac{(AB)}{(A)} > \frac{(\alpha B)}{(\alpha)}$ and negatively associated if $\frac{(AB)}{(A)} > \frac{(\alpha B)}{(\alpha)}$. Putting the given information in a nine-square table.

	A	α	Total
B	3150	2850	6000
β	1350	2650	4000
Total	4500	5500	10000

$$\frac{(AB)}{(A)} = \frac{3150}{4500} = 0.7$$

$$\frac{(\alpha B)}{(\alpha)} = \frac{2850}{5500} = 0.516$$

Since $\frac{(AB)}{(A)}$ is $> \frac{(\alpha B)}{(\alpha)}$ the attributes are positively associated.

(ii) Putting the given information in a nine-square table :

	A	α	Total
B	75	125	200
β	75	125	200
Total	150	250	400

$$\frac{(AB)}{(A)} = \frac{75}{150} = 0.5$$

$$\frac{(\alpha B)}{(\alpha)} = \frac{125}{250} = 0.5$$

Since $\frac{(AB)}{(A)} = \frac{(\alpha B)}{(\alpha)}$, the attributes are independent.

Example 18:

A survey was conducted in respect of marital status and success in examination. Out of 2,000 persons who appeared for an examination, 80% of them were boys, are the test were girls. Among 300 married boys 140 were successful. 1100 boys were successful among unmarried boys. In respect of

100 married girls 40 were successful 200 unmarried girls were successful. Construct two separates nine-square tables, and find out the Yule's Coefficient of Association to discuss the association between marital status and passing of examination.

Solution:

Total no. of persons = 2000

No of boys = 80% = 1600

No of girls = 2000 − 1600 = 400.

Let A denote 'married' and B denote 'successful'. On the basis of given data we will make two separate tables for boys and girls :

Boys : (AB) = 140, (A) = 300, ($\alpha\beta$) = 1100, N = 1600

Putting the information in nine square table :

140	1100	1240
160	200	360
300	1300	1600

$$Q = \frac{(AB)(\alpha\beta) - (A\beta)(\alpha B)}{(AB)(\alpha\beta) + (A\beta)(\alpha B)}$$

$$= \frac{(140 \times 200) - (160 \times 1100)}{(140 \times 200) + (160 \times 1100)}$$

$$= \frac{28000 - 176000}{28000 + 176000} = \frac{148000}{204000} = -0.725.$$

Girls : (AB) = 40, ($\alpha\beta$) = 200, (A) = 100, N = 1600

Putting the information in nine square table :

40	200	240
60	100	160
100	300	400

$$Q = \frac{(AB)(\alpha\beta) - (A\beta)(\alpha B)}{(AB)(\alpha\beta) + (A\beta)(\alpha B)}$$

$$= \frac{(40 \times 100) - (60 \times 200)}{(40 \times 100) + (60 \times 200)}$$

$$= \frac{4000 - 12000}{4000 + 12000} = \frac{-8000}{16000} = -0.5\,.$$

Example 19:

Given in the following information in case of two attributes :

(AB) = 400, (A) = 800, N = 2,500, (B) = 1,600

Name the remaining classes and find out their frequencies.

Solution:

By putting the given information in the nine-square table, we can find out the other frequencies.

	A	α	Total
B	400	1200	1600
β	400	500	900
Total	800	1700	2500

Hence (Aβ) = 400, (αB) = 1200, (αβ) = 500,

(A) = 800, (α) = 1700, (B) = 1600, (β) = 900

Example 20:

Use proportion method to determine the nature of association between A and B :

	B	***b***	***Total***
A	*30*	*50*	*80*
a	*20*	*100*	*120*
Total	*50*	*150*	*200*

Solution:

According to the proportion method if :

$\frac{(AB)}{(A)} = \frac{(\alpha B)}{(\alpha)}$ the attributes A and B are independent.

We have (AB) = 30, (A) = 80, (aβ) = 20, (a) = 120.

$$\frac{(AB)}{(A)} = \frac{30}{80} = 0.375$$

$$\frac{(\alpha B)}{(\alpha)} = \frac{20}{120} = 0.167$$

Since $\frac{(AB)}{(A)} > \frac{(\alpha B)}{(\alpha)}$ the attributes A and B are positively associated.

Example 21:

In an examination at which 500 candidates appeared, boys outnumbered girls by 14 percent of all candidates. Number of passed candidates exceeded the number of failed candidates by 300. Boys failing in examination numbered 80. Construct the nine-square table and calculate the coefficient of association between boys and success in the examination.

Solution:

Let A denote boys, α will denote girls

Let B denote those who passed, β will denote those who failed.

In the information provided we are given

$N = 500, (A\beta) = 80.$

Let x denote girls, boys would be x + 70 (14% of 500)

$x + x + 70 = 500$ [boys + girls = total]

$2x = 500 - 70$

or $x = \frac{430}{2} = 215.$

Hence girls $(\alpha) = 215$ and boys

$(A) = 500 - 215 = 285.$

Let x denote failed candidates, therefore, passed candidates = 300 + x.

Total candidates = 500

Hence, $x + 300 + x = 500$

$2x = 500 - 300$ or $x = 100$

Substituting the given information in a nine-square table,

	A	α	Total
B	205	195	400
β	80	20	100
Total	285	215	500

$$Q = \frac{(AB)(\alpha\beta) - (A\beta)(\alpha B)}{(AB)(\alpha\beta) + (A\beta)(\alpha B)}$$

$$= \frac{(205 \times 20) - (80 \times 195)}{(205 \times 20) + (80 \times 195)}$$

$$= \frac{4100 - 15600}{4100 + 15600} = \frac{11500}{19700} = -0.584.$$

Example 22:

Out of 3,000 unskilled workers of a factory, 2000 come from rural areas and out of 1200 skilled workers, 300 come from rural areas. Determine the association between skill and residence by the method of proportions.

Solution:

Let A denote skilled workers

∴ α will denote unskilled workers

Let B denote workers from rural areas

∴ β will denote workers from urban areas

We are given :

(A) = 1200, (α) = 3000, (B) = 2000, (αβ) = 2000, (AB) = 300

According to the method of proportions, two attributes A and B are said to be independent if :

$$\frac{(AB)}{(A)} = \frac{(\alpha B)}{(\alpha)}$$

In the given case : $\frac{(AB)}{(A)} = \frac{300}{1200} = 0.25;$

$$\frac{(\alpha B)}{(\alpha)} = \frac{2000}{3000} = 0.67$$

Since $\frac{(AB)}{(A)}$ is less than $\frac{(\alpha B)}{(\alpha)}$, there is negative association between skill and residence.

Example 23:

In a class-test in which 135 candidates were examined for proficiency in English and Economics, if was discovered that 75 students failed in English, 80 failed in Economics, and 50 failed in both. Find if there is any association between failing in English and Economics and also state the magnitude of association.

Solution:

Let A denote those who failed in English.

∴ α will denote those who passed in English.

Let B denote failed in Economics.

∴ β will denote those passed in Economics.

Hence given information can be put in a nine-square table and their values determined.

	A	α	Total
B	50	40	90
β	25	20	45
Total	75	60	135

$$Q = \frac{(AB)(\alpha\beta) - (A\beta)(\alpha B)}{(AB)(\alpha\beta) + (A\beta)(\alpha B)}$$

$$= \frac{(50 \times 20) - (25 \times 40)}{(50 \times 20) + (25 \times 40)} = \frac{1000 - 1000}{1000 + 1000} = 0.$$

Example 24:

Show by short-cut method whether there is dissociation or positive or negative association in the following attributes A and B.

(A) = 470; B = 620; (AB) = 320

(AB) = 294; (α) = 570; (αβ) = 380

(αB) = 768; (Aβ) = 480; (αB) = 145.

Solution:

(a) Expectation of (AB) = $\frac{(A) \times (B)}{1000} = \frac{470 \times 620}{1000} = 291.4$

Since (AB) actual frequency is more than (AB) expectation, attributes A and B are positively associated.

(b) We are given (A) and (AB). We need (B) and N

$$N = (A) + (\alpha) = 490 + 570 = 1060$$

$$(B) = (AB) + (\alpha B), (AB) = 249.$$

But (αB) is not given

$$(\alpha B) = (\alpha) - (\alpha\beta) = 570 - 380 = 190$$

Hence $(B) = 294 + 190 = 484$

Expectation of $(AB) = \frac{(A) \times (B)}{N} = \frac{490 \times 484}{1060} = 223.74$

Since (AB) actual frequency is more than (AB) expectation, attributes A and B are positively associated.

(c) We are given (AB) = 256. We have to find (A), (B) and N.

$$N = (AB)\ (A\beta) + (\alpha B) + (\alpha\beta)$$

$$= 256 + 48 + 144 + 768 = 1216$$

$$(A) = (AB) + (A\beta) = 256 + 48 = 304$$

$$(B) = (AB) + (\alpha B) = 256 + 144 = 400$$

Expectation of $(AB) = \dfrac{(A) \times (B)}{N} = \dfrac{304 \times 400}{1216} = 100$

Since (AB) actual frequency is much more than the expectation of (AB), there is positive association between attributes A and B.

Example 25:

A teacher examined 280 students in Economics and Auditing and found that 160 failed in Economics. 140 failed in Auditing and 80 failed in both the subjects. Is there any association between failure in Economics and Auditing?

Solution:

Let A denote students who failed in Economics and B denote students who failed in Auditing.

Putting the given information in a nine-square table, we have

	A	α	Total
B	80	60	140
β	80	60	140
Total	160	120	280

$$Q = \frac{(AB)(\alpha\beta) - (A\beta)\ (\alpha B)}{(AB)(\alpha\beta) + (A\beta)\ (\alpha B)}$$

$$= \frac{(80 \times 60) - (80 \times 60)}{(80 \times 60) + (80 \times 60)}$$

$$= \frac{4800 - 4800}{4800 + 4800} = 0.$$

Since Yule's coefficient of association is zero there is no association between failure in Economics and Auditing.

ASSOCIATION OF THREE ATTRIBUTES

In case of three attributes, the frequencies are grouped as given below :

Order O	N			= 1
Order 1	(A)	(B)	(C)	= 6
	(α)	(β)	(γ)	
Order 2	(AB)	(AC)	(C)	= 12
	(Aβ)	(Aγ)	(Bγ)	
	(αB)	(αC)	(BC)	
	(αβ)	(αγ)	(βγ)	
Order 3	(ABC)	(αBC)		= 8
	(ABγ)	(αBγ)		
	(AβC)	(αβC)		
	(ABγ)	(αβγ)		
				27

Thus in case of three attributes there are 27 distinct frequencies : 1 of order zero, 6 of the first order, 12 of the second order and 8 of the third order.

Any case frequency can always be expressed in terms of class frequencies of higher order. The total number of observations must clearly be equal to the number of A's added to the number α's, *i.e.*,

$$N = (A) + (\alpha)$$

Similarly, the number of A's is equal to the number of A's which are B's added to the number of A's which are B's, *i.e.*

$$(A) = (AB) = (A\beta)$$

Similarly, $(AB) = (ABC) + (AB\gamma)$, and so on.

It follows from the result that any class frequency can be expressed in terms of the frequencies of the highest order, *i.e.*, of order n. For any frequency can be analysed into higher order frequencies, and the process need stop only when we have reached the frequencies of the higher order.

For example, with three attributes

$$(A) = (AB) + (A\beta)$$
$$= (ABC) + (AB\gamma) + (A\beta C) + (A\beta\gamma)$$

The classes specified by n attributes, *i.e.*, those of the highest order, are termed the ultimate class frequencies.

Every class frequency can be expressed as the sum of certain of the ultimate class frequencies. Thus

$$(B) = (ABC) + (AB\gamma) + (\alpha BC) + (\alpha B\gamma) :$$
$$(C) = (ABC) + (\alpha BC) + (A\beta C) + (\alpha\beta C)$$

In case of three attributes the following relationship not be remembered:

$(AB) = (ABC) + (AB\gamma)$ $\quad$ $(BC) = (ABC) + (\alpha BC)$ $\quad$ $(AC) = (ABC) + (A\beta C)$

$(AB) = (ABC) = (A\beta\gamma)$ $\quad$ $(B\gamma) = (AB\gamma) + (\alpha B\gamma)$ $\quad$ $(A\gamma) = (AB\gamma) + (A\beta\gamma)$

$(\alpha B) = (\alpha BC) + (\alpha B\gamma)$ $\quad$ $(\beta C) = (A\beta C) + (\alpha\beta C)$ $\quad$ $(\alpha C) = (\alpha BC) + (\alpha\beta C)$

$(\alpha\beta) = (\alpha\beta C) + (\alpha\beta\gamma)$ $\quad$ $(B) = (BC) + (B\gamma)$ $\quad$ $(C) = (AC) + (\alpha C)$

or $(A) = (AC) + (A\gamma)$ $\quad$ or $(B) = (AB) + (\alpha B)$ $\quad$ or $(C) = (BC) + (\beta C)$

$$N = (ABC) + (AB\gamma) + (A\beta C) + (\alpha BC) + (\alpha B\gamma) + (\alpha\beta\gamma)$$

The following chart will clearly reveal the inter-relationship between frequencies of the different orders :

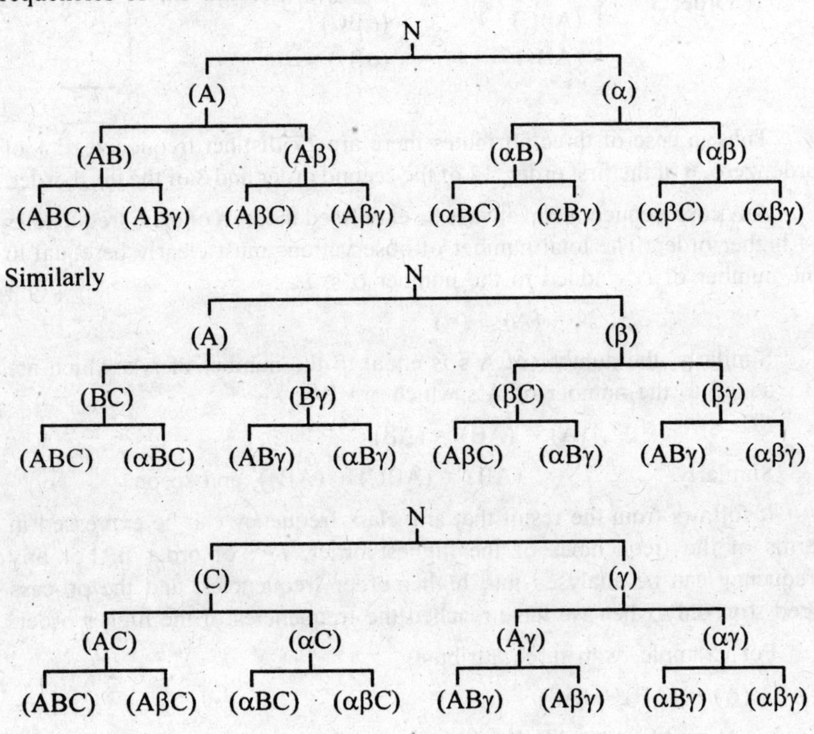

Example 26:

Of 1,000 people consulted, 811 liked chocolates, 752 liked toffees and 418 liked sweets, 570 liked chocolates and toffees; 356 chocolates and sweets and 348 liked toffees and sweets; 297 liked all three. Is this information correct?

Solution:

Denoting liking of chocolates, toffees and sweets by A, B and C respectively, the data would be:

$$N = 1{,}000,\ (A) = 811,\ (B) = 752,\ (C) = 418$$

$$(AB) = 570,\ (AC) = 356,\ (BC) = 348,\ (BC) = 297.$$

If one of the ultimate class frequencies is negative, the given data would be inconsistent.

$$(\alpha BC) = (BC) - (ABC) = 348 - 297 = 51$$

$$(AB\gamma) = (AB) - (ABC) = 570 - 297 = 273$$

$$(ABC) = (AC) - (ABC) = 356 - 297 = 59$$

$$(AB\gamma) = (A) - (AB) - (A\beta C) = 811 - 570 = 152$$

$$(\alpha B\gamma) = (B) - (AB) - (\alpha BC) = 752 - 570 - 51 = 131$$

$$(\alpha\beta C) = (C) - (BC) - (A\beta C) = 418 - 348 - 59 = 11$$

$$(\alpha\beta\gamma) = N - (A) - (B) - (C) + (AB) + (AC) + (BC) - ABC)$$

$$= 1{,}000 - 811 - 752 - 418 + 570 + 356 + 348 - 297 = -4$$

Since $(\alpha\beta\gamma)$ is negative, the given information is not correct.

Example 27:

Given the following frequencies of the positive class, find the frequencies of the rest of the classes :

(A) = 977; (AB) = 453; (ABC) = 127; (B) = 1.185; (AC) = 284;

N = 12,000; (C) = 596; (BC) = 250.

Solution:

In case of three attributes, we have 27 distinct frequencies of which only 8 are given we have to find the remaining 19:

Second Order Frequencies

The following second order frequencies are to be ascertained :

(AB), (αB), $(\alpha\beta)$, $(B\gamma)$, (BC), $(\beta\gamma)$, $(A\gamma)$, (αC), $(\alpha\gamma)$,

$$(A\beta) = (A) - (AB) = 977 - 453 = 524$$

$$(\alpha B) = (B) - (AB) = 1185 - 453 = 732$$

$$(\alpha\beta) = (\alpha) - (\alpha B) = 11023 - 732 = 10{,}291 \quad [\because (\alpha) = N - (A)]$$

$$(B\gamma) = (B) - (BC) = 1185 - 250 = 935$$

$(\beta C) = (C) - (BC) = 596 - 250 = 346$

$(\beta\gamma) = (\beta) - (BC) = 10815 - 336 = 10,469 \quad [\because (\beta) = N - (B)]$

$(A\gamma) = (A) - (AC) = 977 - 284 = 693$

$(\alpha C) = (C) - (AC) = 596 - 248 = 212$

Third Order Frequencies :

The third order frequencies to be obtained are $(AB\gamma)$, (αBC), $(\alpha B\gamma)$, $(\alpha\beta C)$, $(\alpha\beta\gamma)$.

$(AB\gamma) = (AB) - (ABC) = 453 - 127 = 326$

$(\alpha BC) = (BC) - (ABC) = 250 - 127 = 123$

$(\alpha B\gamma) = (\alpha B) - (\alpha BC) = 732 - 123 = 609$

$(A\beta C) = (AC) - (ABC) = 284 - 127 = 157$

$(A\beta\gamma) = (A\beta) - (A\beta C) = 524 - 157 = 367$

$(ABC) = (\beta C) - (A\beta C) = 346 - 157 = 189$

$(\alpha\beta\gamma) = (\alpha\beta) - (\alpha\beta C) = 10291 - 189 = 10102$

It may be noted that there are more than one way in which any one class frequency can be obtained.

Example 28:

The following are the proportions per 5,000 of workers observed for certain classes of defects among a number of factory workers :

A = Development defects; B = Nerve signs; C = Mental dullness.

N = 5,000 (C) = 400; (A) = 440; (AB) = 170; (B) = 545; (BC) = 228

Show that some dull workers do not exhibit development defects and state how many at least do not do so.

Solution:

It is required to find the least value of (αC).

By the condition of consistencies

$$(BC) - (AB) + (AC) \not> (C); (AC) > -(BC) + (AB) + (C)$$

Now $\quad -(BC) + (AB) + (C) = -228 + 170 + 400 = 342$

$\therefore \quad (AC) \not> 342; (C) = (\alpha C) \not> 342$ or $(\alpha C) \not> 400 - 342 = 58$

$\therefore$ At least 58 dull workers do not exhibit development defects.

Example 29:

At a competitive examination at which 600 graduates appeared boys outnumbered girls by 96. Those qualifying for interview exceeded in number those failing to qualify by 310. The number of Science graduate boys interviewed was 300 while among the Arts girls there were 250 who failed to qualify for interview. Altogether there were only 135 Arts graduates and 33 among them failed to qualify. Boys who failed to qualify numbered 18.

Find (a) the number of boys who qualified for interview, (b) the total number of Science graduate boys appearing, and (c) the number of Science graduate girls who qualified.

Solution:

Let us denote boys by A, qualifying for interview by B and offering Science C, the data then, are :

$$N = 600 \qquad (\alpha\beta\gamma) = 25$$
$$(A) - (\alpha) = 96 \qquad (\gamma) = 135$$
$$(B) - (\beta) = 310 \qquad (\beta\gamma) = 35$$
$$(ABC) = 300 \qquad (AB) = 18$$

We are to find the value of (i) (AB); (ii) (AC) and (iii) (αBC)

$$N = (A) + (\alpha) = 600;$$
$$(A) - (a) = 96$$
$$2(A) = 696 \text{ or } (A) = 348$$
$$(B) + (\beta) = 600; (B) - (\beta) = 310$$
$$2(B) = 910 \text{ or } (B) = 455 \text{ and } (\beta) = 145$$
$$(C) = N - (\gamma) = 600 - 135 = 465$$

Hence the number of boys who qualified for interview was 330:

$$(AC) = (ABC) + (A\beta C) = (ABC) = (A\beta) - (AB\gamma)$$
$$= (ABC) + (A\beta) - [(\beta\gamma) - (\alpha\beta\gamma)] = 300 + 18 - 33 + 25 = 310$$

Hence the total number of Science graduate boys appearing was 310:

$$(\alpha BC) = (BC) - (ABC) = (C) = (\beta C) = (ABC)$$
$$= (C) - [\beta - (\beta\gamma)] - (ABC) = 465 - 145 + 33 - 300 = 53.$$

Hence the number of Science graduate girls who qualified was 53.

Consistency of Data : For three attributes the following conditions should be satisfied otherwise the frequencies given on the right will be negative:

(a) (ABC) < O otherwise (ABC) will be negative

(b) (ABC) < (AB) + (AC) – (A) otherwise (Aβγ) will be negative

(c) (ABC) < (AB) + (BC) – (B) otherwise (αBγ) will be negative

(d) (ABC) < (AC) + (BC) – (C) otherwise (αβC) will be negative

(e) (ABC) > (AB) otherwise (ABγ) will be negative

(f) (ABC) > (AC) otherwise (AβC) will be negative

(g) (ABC) > (BC) otherwise (αBC) will be negative

(h) (ABC) > (AB) + (AC) + (BC) otherwise (ABC) will be negative
– (A) – (B) – (C) + N otherwise (αBC) will be negative

Since all these comparisons are not independent, we get the following four conditions : (i) is obtained by combining (a) and (h), (f) by (b) and (g) and so on.

(i) (AB) + (AC) + (BC) $\nless$ (A) + (B) + (C) – (N)

(j) (AB) + (AC) – (BC) $\ngtr$ (A)

(k) (AB) – (AC) + (BC) $\ngtr$ (B)

(l) (AC) + (NC) – (AB) $\ngtr$ (C)

Or

(AB) + (AB) – (A) $\ngtr$ (BC)

(AB) + (BC) – (B) $\ngtr$ (AC)

(AC) + (BC) – (C) $\ngtr$ (AB)

Example 30:

100 children took three examinations 40 passed the first, 39 passed the second and 48 passed the third, 19 passed all three, 9 passed first two and failed in third, 19 failed in the first two and passed the third. Find how many children passed at least two examinations. Show that for the question asked certain of the given frequencies are not necessary.

Solution:

We are given

(A) = 40, (B) = – 39, (C) = 48, (ABC) = 19,

(ABγ) = 9, (αβC) = 19, N = 100

We have to find the number of children who passed at least two examinations.

i.e. (ABC) + (ABγ) + (AγC) + (αBC) + (AβC). Of these we are given (ABC) and (ABγ).

We have to find (AβC) and (αBC).

$$(C) = (AC) + (\alpha C) = (ABC) + (A\beta C) + (\alpha BC) + (\alpha\beta C)$$

$$48 = (ABC) + (A\beta C) + (\alpha BC) + 19$$

$$\therefore (ABC) + (A\beta C) + (ABC) = 48 - 19 = 29.$$

$$\therefore (ABC) + (A\beta C) + (\alpha BC) + (AB\gamma) = 29 + 9 = 38.$$

If should be noted that we required only (C), (αβC); the other frequencies are not necessary.

Example 31:

In a very hotly fought battle,

70% at least of combatants lost an eye.

75% at least an ear.

80% at least a leg.

85% at least an arm.

What percentage at least lost all the four organs?

Solution:

Let A stand for combatants who lost an eye, B for those who lost an ear, C for those who lost a leg and D for those who lost an arm.

N stands for the total number of combatants.

Then (A) = 0.70N, (B) = 0.75N, (C) = 0.80N, (D) = 0.85N

We have to find the least value of (ABCD)

$$(ABCD) = (A) + (B) + (C) + (D) - 3N$$

$$= 0.70N - 0.75N + 0.80N + 0.85N - 3N = 0.10N$$

The least value of (ABCD) = 0.10N or 10%

Hence 10% at least of combatants lost for the four organs.

Example 32:

Given the following data, find frequencies of (i) the remaining positive classes, and (ii) ultimate classes.

$$N = 1{,}800,\ (A) = 850,\ (B) = 780,\ (C) = 326.$$

$$(ABg) = 200,\ (AbC) = 94,\ (aBC) = 72,\ (ABC) = 50.$$

Solution:

We are to find out (i) (AB), (AC) and (BC),

(ii) (Aβg), (αBγ), (αβC), (αβγ).

$(AB) = (ABC) + (AB\gamma) = 50 + 200 = 250$

$(BC) = (ABC) + (\alpha BC) = 50 + 72 = 122$

$(AC) = (ABC) + (A\beta C) = 50 + 94 = 144$

$(A\beta\gamma) = (A\beta) - (A\beta C)$

or $(A) - (AB) - (A\beta C) = 850 - 250 - 84 = 506$

$(\alpha B\gamma) = (\alpha B) - (\alpha BC) = (B) - (AB) = (\alpha BC)$

$= 780 - 250 - 71 = 458$

$(\alpha\beta C) = (\beta C) - (A\beta C) = (C) - (BC) - (A\beta C)$

$= 320 - 111 - 23 = 110$

$(\alpha\beta\gamma) = N - (A) - (B) - (C) + (AB) + (BC) + (AC) - (ABC)$

$= 1100 - 820 - 780 - 326 + 250 + 112 + 144 + 50 = 310$

Example 33:

There were 400 students in B.Com. (Hons) in a certain college. The results in various terminal examinations are given below :

180 passe din first terminal

140 passed in second terminal

180 passed in third terminal

60 passed in all terminals

80 failed in all terminals

40 passed in the first and second terminals but failed in the third terminal, 70 failed in the first and second terminals but passed in the third terminal.

Find out how many students passed at least two examination.

Solution:

Let success in first terminal be denoted by A. Success in second terminal by B, and

Success in third terminal by C.

α, β and γ will denote failure in 1st, 2nd and 3rd terminals, respectively.

We are given N = 400, (A) = 180, (B) = 140, (C) = 180, (ABC) = 6, $(\alpha\beta\gamma) = 80$, $(AB\gamma) = 40$, $(\alpha\beta C) = 70$.

We have to calculate $(ABC) + (AB\gamma) + (\alpha\beta C) + (A\beta C)$ to get the required answer.

$(ABC) = 60$, $(AB\gamma) = 40$ (given)

$(C) = (ABC) + (\alpha\beta C) + (\alpha BC)$

$180 = 60 + 70 + (A\beta C) + (\alpha BC)$

$(\alpha BC) + (\alpha\beta C) = 180 - 60 - 70 = 50$

$\therefore (ABC) + (AB\gamma) + (\alpha BC) + (A\beta C) = 60 + 40 + 50 = 150.$

i.e., 150 students passed at lest two examinations.

Example 34:

Do you find any inconsistency in the data given below :

$N = 1000,\ (A\beta) = 483,\ (A\gamma) = 378,\ (B\gamma) = 226$

$(A) = 525,\ (B) = 312,\ (C) = 470 \text{ and } (ABC) = 25$

Solution:

This is a case of three attributes. The ultimate class frequencies are $2^3 = 8$ in number and are as follows :

$(ABC), (AB\gamma), (A\beta C), (AB\gamma), (\alpha BC), (\alpha\beta B), (\alpha\beta\gamma)$

Before proceeding to evaluate them we find the value of

$(AB) = (A) - (A\beta) = 525 - 483 = 42$

$(AC) = (A) - (A\gamma) = 525 - 378 = 147$

$(BC) = (B) - (B\gamma) = 312 - 226 = 86$

Now the ultimate frequencies are :

$(ABC) = 25$ (given)

$(A\beta\gamma) = (AB) - (ABC) = 42 - 25 = 17$

$(A\beta C) = (AC) - (ABC) = 147 - 25 = 122$

$(A\beta\gamma) = (A\beta) - (A\beta C) = 86 - 25 = 61$

$(\alpha BC) = (BC) - (ABC) = 86 - 25 = 61$

$(\alpha B\gamma) = (\alpha B) - (\alpha BC) = (B) - (AB) - (AB) - (\alpha BC)$

$= 312 - 42 - 61 = 209$

$(\alpha\beta C) = (C) - (AC) - (BC) + (ABC)$

$= 470 - 147 - 86 + 25 = 262$

$(\alpha\beta\gamma) = N - (A) - (B) - (C) + (AB) + (AC) + (BC) - (ABC)$

$= 1000 - 525 - 312 - 470 + 42 + 147 + 86 - 25 = -57$

As one of the ultimate class frequencies is negative, the given data are inconsistent.

PARTIAL ASSOCIATION

So far we have discussed the association of A and B in the universe as a whole without finding the other attributes in the universe. It is possible, however, that there is no direct relationship between A and B, *i.e.*, the association between attributes A and B may be due to their association with a third attribute say, C. Thus if A is positively associated with C and if B is associated with C. A may be found to be positively associated with B. But this type of association between A and B is not direct—It is the effect of their association with a third attribute C. To find out whether the association with a third attribute C. It would be necessary to study the association of A and B in the sub-populations C and γ. If A and B are associated in both the sub-populations of C and γ it would indicate that A and B are really associated with each other.

The association between A and B in sub-populations are called partial associations, to distinguish them from the total associations between A and B in the population at large. The following example will illustrate clearly the concept of partial association :

An association is observed between calculations and prevention from attack by small-pox. It means that vaccination prevents attack of small-pox. However, on a detailed analysis one may find that the attributes vaccination and attack of small-pox are not directly associated—the association between them is due to a third factor, namely economic condition. Those, people who are economically well-off live in better conditions, open house, get better food and nutrition and can afford the cost of vaccination and as such the possibility of their getting an attack of small-pox is less. On the other hand, those who are poor, live in filthy conditions, dirty surroundings, dirty houses and because of illiteracy do not believe in vaccination. They cannot afford the cost of vaccination also. As such they are liable to suffer more from diseases. If we denote A for vaccination. B for small-pox and C for economic conditions, we may find that there is positive association between A and C and also between B and C. Hence in order to arrive at correct conclusions it is necessary that on the basis of economic conditions the population is divided into two parts rich (C) and poor (γ) and in each sub-population association is ascertained between vaccination (A) and prevention from small-pox (B). If this third attribute is ignored it will give rise to misleading conclusions or technically illusory association.

The association found between the attributes A and B in the universe of C's and universe of γ's is termed as partial association to distinguish them from total association found between A and B in the universe at large.

The methods of finding out partial association are the same as used in finding the total association. The only difference is that we have to find out separately the association of A and B in C and γ.

First Method

Association between A and B for sub-population	Independent	Positive Association	Negative Association
C	$(ABC) = \frac{(AC)\times(BC)}{(C)}$	$(ABC) > \frac{(AC)\times(BC)}{(C)}$	$(ABC) < \frac{(AC)\times(BC)}{(C)}$
γ	$(AB\gamma) = \frac{(A\gamma)\times(B\gamma)}{(C)}$	$(AB\gamma) > \frac{(A\gamma)\times(B\gamma)}{(C)}$	$(AB\gamma) < \frac{(A\gamma)\times(B\gamma)}{(C)}$

Second Method

C	$\frac{(ABC)}{(AC)} = \frac{(\alpha BC)}{(\alpha C)}$	$\frac{(ABC)}{(AC)} > \frac{(\alpha BC)}{(\alpha C)}$	$\frac{(ABC)}{(AC)} < \frac{(\alpha BC)}{(\alpha C)}$
γ	$\frac{(AB\gamma)}{(A\gamma)} = \frac{(\alpha B\gamma)}{(\alpha\gamma)}$	$\frac{(AB\gamma)}{(A\gamma)} > \frac{(\alpha B\gamma)}{(\alpha\gamma)}$	$\frac{(AB\gamma)}{(A\gamma)} < \frac{(\alpha B\gamma)}{(\alpha\gamma)}$

Third Method (Yule's Coefficient)

$$Q_{AB.C} = \frac{(ABC)(\alpha\beta C) - (A\beta C)(\alpha BC)}{(ABC)(\alpha\beta C) + (A\beta C)(\alpha BC)}$$

$Q_{AB.C}$ = Coefficient of partial association between A and B for sub-population C

$$Q_{AB.\gamma} = \frac{(AB\gamma)(\alpha\beta\gamma) - (A\beta\gamma)(\alpha B\gamma)}{(AB\gamma)(\alpha\beta\gamma) + (A\beta\gamma)(\alpha B\gamma)}$$

Association between A and C for sub-population B

$$Q_{AC.B} = \frac{(ABC)(\alpha\beta\gamma) - (A\beta\gamma)(\alpha BC)}{(ABC)(\alpha\beta\gamma) + (A\beta\gamma)(\alpha BC)}$$

$$Q_{AC.B} = \frac{(A\beta C)(\alpha\beta\gamma) - (A\beta\gamma)(\alpha\beta C)}{(A\beta C)(\alpha\beta\gamma) + (A\beta\gamma)(\alpha\beta C)}$$

ILLUSORY ASSOCIATION

Although a set of attributes independent of A and B will affect the association between them the existence of an attribute C with which they are both associated may give an association in the population at large which is illusory in the sense that it does not correspondent to any real relationship between them.

Illusory association may also arise in a different way through the personality of the observer or observers. If the observer's attention fluctuates, he may be more likely to notice the presence of A when he notices the presence of B and vice versa. In such a case A and B (so far as the record goes) will both be association will be created. Again, if the attributes are not well defined, one observer may be more generous than another in deciding when to record the presence of A and also the presence of B and even one observer may fluctuate in the generosity of his marking. In this case the recording of A and the recording of B will both be associated with the generosity of the observer in recording their presence, C, and an illusory association between A and B will consequently arise.

LIST OF FORMULAES

1. *Comparison of Observed and Expected Frequency Method*

   Attributes A and B are said to be :

   (i) Independent if $(AB) = \dfrac{(A) \times (B)}{N}$

   (ii) Positively associated if, $(AB) > \dfrac{(A) \times (B)}{N}$

   (iii) Negatively associated if, $(AB) < \dfrac{(A) \times (B)}{N}$

2. *Proportion Method*

   Attributes A and B are said to be

   (i) Independent if, $\dfrac{(AB)}{(A)} = \dfrac{(\alpha B)}{(\alpha)}$

   (ii) Positively associated if, $\dfrac{(AB)}{(A)} > \dfrac{(\alpha B)}{(\alpha)}$

   (iii) Negatively associated if, $\dfrac{(AB)}{(A)} < \dfrac{(\alpha B)}{(\alpha)}$

3. *Yule's Coefficient of Association*

$$Q = \frac{(AB)(\alpha\beta) - (A\beta)(\alpha B)}{(AB)(\alpha\beta) + (A\beta)(\alpha B)}$$

4. *Coefficient of Colligation*

$$Y = \frac{1 - \sqrt{\dfrac{(A\beta) \times (\alpha B)}{(AB) \times (\alpha\beta)}}}{1 + \sqrt{\dfrac{(A\beta) \times (\alpha B)}{(AB) \times (\alpha\beta)}}}$$

Coefficient of Contingency

$$C = \sqrt{\frac{\chi^2}{N + \chi^2}}$$

5. *Spearman's Rank Correlation Coefficient*

$$R = 1 - \frac{6\Sigma D^2}{N^3 - N}$$

In case ranks are repeated

$$R = 1 - \frac{6\left(\Sigma D^2 + \frac{1}{12}(m^2 - m) + \frac{1}{12}(m^3 - m)\right)}{N^3 - N}$$

6. *Concurrent Deviation Method*

$$r_C = \pm\sqrt{\left(\frac{2C - n}{n}\right)}$$

$$S.E._r = \frac{1 - r^2}{\sqrt{N}};\ P.E._r = 0.6745\frac{1 - r^2}{\sqrt{N}}$$

Coefficient of determination = r^2.

EXERCISES

1. Explain briefly the method of finding association between attributes.
2. What is meant by Association of two attributes? Explain briefly the different methods of measuring association between two attributes.
3. The following information is available in respect of two attributes A and B :

	A	*(α)*
B	14	110
β	16	20

Find Yule's coefficient of association between attributes (i) 'A' and β, (ii) α and β.

[Q = 0.725]

4. In a market survey of 10000 persons, it was that 8110 liked chocolates, 7520 liked toffee, 4180 liked bottled sweets, 5700 liked chocolates and toffee, 350 liked chocolate and boiled sweets, 3480 liked toffee and boiled sweets and 2970 liked all the three. Show that this information, as it stands, must be incorrect.
5. What do you mean by the coefficient of contingency? When does it assume maximum value.
6. In a report on consumers preference, it was given that out of 500 persons surveyed 410 preferred variety A. 380 preferred variety B and 270 persons liked both. Are the data consistent?

   [(αβ) = – 20; no]
7. Explain from the data given below whether A and B are independent, positively associated or negatively associated :

   (i) N = 10,000; (A) = 4,500; (B) = 6,000; (AB) = 3,150.

   (ii) (A) = 150; (α) = 250; (B) = 200; (αβ) = 125.

   (iii) N = 2,500; (AB) = 600, (α) = 750; (β) = 1,500.

   [(i) Positive (ii) Negative (iii) Negative]
8. What do you understand by consistency of data? Illustrate with examples.
9. Write down the conditions under which two attributes A and B are (i) independent (ii) positively associated and (iii) negatively associated. Explain the conditions.
10. 200 candidates appeared for a competitive examination and of these 60 were successful, 35 had attended a coaching class and of these 20 come out to be successful. Estimate the utility of the coaching class.

    [Q = 0.612]
11. Find out the coefficient of association between the type of college training and success in teaching from the following table :

Teachers	*Successful*	*Unsuccessful*	*Total*
College	75	55	130
University	125	45	170
Total	200	100	300

Also calculate coefficient of contingency.

[Q = – 0.341, C = 0.025]

12. Calculate coefficient of contingency from the following data :

Social status	*Dull*	*Intelligent*	*Brilliant*
Lower Middle	22	35	23
Middle	38	70	32
Upper Middle	60	20	20

[Q = 0.315]

13. (a) (i) Determine the consistency or inconsistency of the following data :

(AB) = 200, N = 1000, (A) = 150, (B) = 300.

(ii) In a co-educational institution, out of 200 students, 150 were boys. In an examination, 120 boys and 40 girls passed. Apply Yule's co-efficient to determine the association between sex and success in the examination. Interpret your result.

(b) In a survey, the following data are available :

	First Son	
	Literature	*Illiterate*
Literate Father	60	40
Illiterate :	40	20

Find whether the literacy of first son has association with the literacy of the father.

[Q = – 0.143] (B.A. Bharathidasan, 1997)

14. Calculate Yule's coefficient of association between literacy and unemployment from the data given below :

Total Adults	:	2,25,000
Literates	:	50,000
Unemployed	:	17,000
Literate and unemployed	:	7,000

[Q = 0.457]

15. Investigate the association between darkness of eye colour in fathers and sons from the following data :

Combination	*Frequency*
Fathers with dark eyes and sons with dark eyes	50
Fathers with dark eyes and sons with not dark eyes	79
Fathers with not dark eyes and sons with dark eyes	89
Fathers with not dark eyes and sons with not dark eyes	82

16. In an analysis of two attributes, if N = 160, (A) = 96 and (B) = 50, find the frequencies of the remaining classes on the assumption that A and B are independent.

17. From the adult population of our large cities, random samples of sizes given below, were taken and the numbers of married and single mess recorded. Do the data indicate any association between cities and marriage?

City	*A*	*B*	*C*	*D*
Married	137	164	152	
Single	32	57	56	
Total	169	221	208	

18. (a) What is meant by Association of two Attributes? How is it measured?

    (b) How would you distinguish between association and correlation as the terms are used in statistics?

19. (a) Explain the terms 'Association and Disassociation' between two attributes with examples.

    (b) Define Coefficient of Association. What are its properties?

20. How will you examine the consistency of data classified according to different attributes? Give a set of conditions of consistencies in case of two attributes.

21. When are two attributes said to be :

    (a) independent

    (b) positively associated, and

    (c) negatively associated?

    What are the conditions to be satisfied by class frequencies in each of the above cases?

22. Do you find any association between the temperaments of brothers and sisters from the following data :

Good-natured brothers and good-natured sisters	1230
Good-natured brothers and sullen sisters	850
Sullen brothers and good-natured sisters	530
Sullen brothers and sullen sisters	980

[Q = + 0.456]

23. Find the Association between Literacy and Unemployment from the following figures :

Total Adults	10,000
Literates	1,290
Unemployed	1,390
Literate Unemployed	7,820

[Q = 0.923]

24. Which of the following statements are True and False :

(i) If the frequencies are grouped into a table with three rows and three columns, it is called 2 × 3 contingency table T/F

(ii) If Q = ± 1, there is perfect positive association between the attributes. T/F

(iii) Two attributes, A and B are said to be positively associated if:

$$(AB) < \frac{(A) \times (B)}{N}$$

(iv) If (B) = 500 and (N) = 700, (β) shall be 500+700=1,200. T/F

(v) The formula for calculating coefficient of association was given by Karl Persons. T/F

(vi) For three attributes, A, B, C the number of ultimate class frequency is $2^3 = 8$. T/F

(vii) If any of the ultimate class frequencies is negative, the given data are called inconsistent. T/F

[**Ans.** (i) F (ii) T (iii) F (iv) F (v) F (vi) T (vii) T]

25. According to a survey, the following results were obtained :

No. of coordinates appeared at the exam. = 800

Married	:	150
Married & Successful	:	70
Unmarried & Successful	:	550

Find the association between married status and the success in the examination.

[Q = – 0.725]

26. In order to ascertain if the marriage has any effect on the examination results of students. 1,000 students were selected at random. There were Hindus. Muslims & Christians of the 1000 students, 375 were married. Of the married students, 167 passed and of the unmarried

students. 203 failed. Find Yule's coefficient of association between marriage and failure of students in the examination.

27. Tick the correct answer :

(a) Attributes A and B are said to be positively associated, if :

(i) $(AB) = \frac{(A) \times (B)}{N}$

(ii) $(AB) < \frac{(A) \times (B)}{N}$

(iii) $(AB) < \frac{(A) \times (B)}{N}$

(iv) $(AB) = \frac{(A) \times (B)}{N}$

(v) $(AB) = \frac{(A) \times (B)}{N}$.

(b) Yule's coefficient of association can be :

(i) greater than 1

(ii) less than 1

(iii) cannot be zero

(iv) has no limits

(v) can take any value between ± 1.

(c) Coefficient of contingency, *i.e.* C is calculated as :

(i) $\sqrt{\frac{\chi^2}{N}}$ (ii) $\frac{\sqrt{\chi^2}}{N}$

(iii) $\frac{\chi^2}{N + \chi^2}$ (iv) $\sqrt{\frac{\chi^2}{N + \chi^2}}$

(v) $\frac{\chi^2}{\sqrt{N + \chi^2}}$

(d) When there is perfect positive association between attributes, Q would be :

(i) Zero; (ii) – 1

(iii) – 0.9; (iv) none of these.

(e) In case of three attributes the number of distinct frequencies is:

(i) 12; (ii) 27; (iii) 15;

(iv) 25; (v) none of these.

(f) If N = 500, (B) = 300, (B) shall be :

(i) 200; (ii) 800

(iii) 1,000; (iv) none of these.

[Ans. (a) (iii), (b) (v), (c) (iv), (d) (iv), (e) (ii), (f) (i)]

28. State the criteria for consistency of data collected for studying association of two attributes.

29. You are given the following information :

(A) = 400, (AB) = 250, (B) = 550, N = 1,200

Prepare the nine-square table and find the missing frequencies.

[(Aβ) = 150 ($\alpha\beta$) = 250 ($\alpha\beta$) = 550, (β) = 700, (α) 800]

30. Explain the concept of (i) complete association, (ii) partial association, (iii) complete disassociation, and (iv) independence, so as to make the difference between them clear.

31. What do you understand by Illusory Association? Explain with the help of an example.

32. Describe the conditions for consistency of data when we are dealing with three attributes A, B and C.

33. Out of 700 literates in a particular taluk, number of criminals was 5. Out of 9,300 illiterates in the same taluk, the number of criminals was 150.

On the basis of these figures, find Yule's coefficient of association between illiteracy and criminality.

[Q = – 0.39]

34. Find all the ultimate class-frequencies from the following data :

N = 800, (A) = 224, (B) = 301, (C) = 150, (AB) = 125, (AC) = 72, (BC) = 60, (ABC) = 32.

35. The following information was provided by an investigator who interviewed 1,000 students. 811 read Hindustan Times, 752 read Statesman and 418 read Times of India; 570 read Hindustan Times and Stateman, 356 read Hindustan Times and Times of India and 348 read Statesman and Times of India : 279 read all the three.

Is there any inconsistency in the above information?

36 What do you understand by association of attributes? How will you examine the consistency of data classified according to different attributes?

37. Find out association between the intelligence of husbands and the intelligence of wives from the following data :

Intelligent husbands with intelligent wives	40
Intelligent husbands with dull wives	100
Dull husbands with intelligent wives	160
Dull husbands with dull wives	190

[Q = – 0.397).

38. From the table given below, calculate Yule's coefficient of association between literacy and unemployment in urban and rural areas :

Area	*Total*	*Literate*	*Unemployment*	*Literate unemployment*
Urban (Lakhs) :	25	10	5	3
Rural (Lakhs) :	200	40	12	4.

39. What do you understand by 'Association of Attributes'? Differentiate between association, disassociation and independence. Describe the most commonly used methods for determining association between two attributes.

40. From the following data, relating to 50 persons, find Yule's coefficient of association and interpret it :

Illiterate Honest	:	22
Literate Honest	:	2
Illiterate Dishonest	:	18

41. Discuss the concept of association between two attributes and correlation between two variables.

42. From the following given frequencies, find out all the remaining class frequencies :

N = 1,000, (A) = 88, (B) = 109, (C) = 28
(AB) = 34, (AC) = 14, (BC) = 13, (ABC) = 6.

43. Examine the consistency of the following data regarding two attributes.

N = 600, (α) = 350, (B) = 300, (Aβ) = 150.

44. Explain the concept of Association and Independence of attributes as applied to a 2 × 2 contingency table. Discuss the different coefficients proposed for measuring association. Use one of them for measuring

the association between proficiency in English and in Hindi among candidates at a certain test if 245 of them passed in Hindi, 285 failed in Hindi, 190 failed in Hindi but passed in English, and 147 passed in both.

[Q = – 0.143].

45. Using the following data, find Yule's coefficient of colligation and explain the result.

$$(A) = 120,\ (B) = 75,\ (\alpha\beta) = 55,\ N = 200.$$

46. A college submitted the following returns to university office :

Number of students appearing in all the three tests	= 350
Number of students appearing passing the first test	= 125
Number of students appearing passing the second test	= 135
Number of students appearing passing the third test	= 145
Number of students appearing passing all the three tests	= 20
Number of students appearing failing in all the three tests	= 75
Number of students appearing passing the first two but failing in the third	= 25
Number of students appearing failing in the first two but passing in the third	= 60

From the information given above, find out the number of students passing at least two tests.

47. Examine whether the following data are consistent. If so, find Yule's coefficient of colligation.

$$N = 200,\ (AB) = 24,\ (\alpha) = 160,\ (\alpha\beta) = 70$$

48. (a) Explain Yule's coefficient of association.

49. Fill in the blanks :

(i) Association of attributes helps us to study the relationship between phenomena which are of .... nature.

(ii) (AB) denotes the number of individuals possessing attributes ....

(iii) When we study two attributes, the total frequencies are ....

(iv) If $\frac{(A)\times(B)}{N} = (AB)$, the attributes are called ....

(v) The order of classes depends upon the number of .... under study.

(vi) The Yule's Coefficient of association varies between ....

(vii) If (A) = 200 and (B) = 70, (AB) shall be ....

[Ans. (i) qualitative (ii) A and B (iii) 9 (iv) independent (v) attributes (vi) + 1 (vii) 130]

50. (a) Examine whether A and B are independent in the following case:

$N = 5,000, (A) = 2,350, (B) = 3,104, (AB) = 1,600$

(b) Find whether A and B are independent in the following case :

$(AB) = 256, (\alpha\beta) = 768, (A\beta) = 48, (\alpha\beta) = 144.$

51. In a population of 1,000 students, the number of married is 400. Out of the 300 students who failed, 120 belonged to the married group. Using Yule's coefficient of association, find out the extent of association of the attributes, marriage and failure.

52. From the following table, discuss whether the colour of the sons eyes is associated with that of the father or not?

	Eye Colour (Sons)	
Eye Colour (Fathers)	*Not Light*	*Light*
Not Light	230	148
Light	151	147

[Q = – 0.63]

53. (a) Distinguish between concepts of association and correlation and describe the situations in which each of them should be used. Illustrate your answer with examples.

(b) When are two attributes A and B, each occurring at two levels, said to be independent? When are they said to be associated?

(c) What do you understand by association of attributes and coefficient of association?

54. (a) What is meant by association. Distinguish between 'Association' and 'Correlation' and explain the theory and measurement of association of attributes.

55. (a) Explain a contingency table with an example.

(b) Write down the two formulae for finding the association between two attributes.

(c) What is a contingency table? Explain the method of its construction.

56. (a) Calculate Yule's coefficient of association between smoking and coffee drinking habits :

Habits	*Coffee drinkers*	*Non-coffee drinkers*
Smokers	90	65
Non-smokers	260	110

(b) If (AB) = 50, (Aβ) = 79, (αB) = 89, (αβ) = 782, find Q_{AB}.

[(a) Q = – 0.261, (b) Q = 0.695]

57. In a diphtheria epidemic out of 508 children who were not protected against it, 40 suffered and 24 of them died. But out of 588 children to whom triple antigen was injected, 12 suffered and only 2 died. Do you find any association between :

    (i) Inoculation and suffering from diphtheria.

    (ii) Inoculation and mortality among children who suffered from it?

58. In the context of dichotomous classification with three attributes, examine whether the following data are consistent :

    N = 1,000 (A) = 510, (B) = 490, (C) = 427,

    (AB) = 189, (AC) = 140, (BC) = 85.

59. From the data given below, compare the association between literacy and unemployment in the rural and urban areas and given reasons for the difference, if any :

		Rural	*Urban*
Total number of adult males	:	25 lakhs	200 lakhs
Literate males	:	10 lakhs	40 lakhs
Unemployed males	:	5 lakhs	12 lakhs
Literate and unemployed males	:	3 lakhs	4 lakhs

[Q. (Urban) = 0.472; (Rural) = 0.357]

60. The following table gives the number of literates and criminals in three cities. Compare the degree of association between literacy and criminality in these cities :

	Jalandhar	*Jammu*	*Jodhpur*
Total No. ('000)	244	184	230
Literates ('000)	40	47	33
Literate criminals ('000)	3	3	2
Illiterate criminals ('000)	40	20	24

[Jalandhar = + 0.45, Jammu = 0.55, Jodhpur = + 0.34]

61. In a sample study about the coffee habits in two towns, following data were observed :

Town X : 50% persons were male, 30% were coffee drinkers and 18% were male coffee drinkers.

Town Y : 45% were males, 25% were coffee drinkers and 16% were male coffee drinkers. Is there any association between sex and coffee habits? If so, in which town it is greater?

62. Calculate the coefficient of association between extravagance in fathers and sons from the following data :

Extravagant fathers with Extravagant sons	327
Extravagant fathers with Miserly sons	545
Miserly fathers with Extravagant sons	741
Miserly fathers with Miserly sons	235

[Q = – 0.68].

63. Distinguish between Association of Attributes and Correlation.

4

Statistical Inference—Tests of Hypothesis

INTRODUCTION

It is a branch of statistics which deal with uncertainty in decision-making.

It refers to the process of selecting and using a sample statistic to draw inference about a population parameter based on a subset of it– the sample drawn from the population. Statistical inference treats two different classes of problems :

1. *Hypothesis testing, i.e.*, to test some hypothesis about parent population from which the sample is drawn.
2. *Estimation, i.e.,* to use the 'statistics' obtained from the sample as estimate of the unknown 'parameter' of the population from which the is drawn.

HYPOTHESIS TESTING

A hypothesis is a supposition made as a basis for reasoning. According to Prof. Morris Hamburg. "A hypothesis in statistics is simply a quantitative statement about a population." Palmer O Johnson has beautifully described hypothesis as "islands in the uncharted seas of thought to be used as bases for consolidation and recuperation as we advance into the unknown."

The method of testing hypothesis is given as under :

(1) *Set up a Hypothesis* : The test of the hypothesis we first setup a hypothesis about a population parameter. Then we collect some sample data, produce sample statistics, and use this information to decide how likely it is that our hypothesized population parameter is correct. Say, we assume a certain value for a population mean. To test the validity of our assumption, we gather sample data and determine the difference between the hypothesized value and the

actual value of the sample mean. Then we judge whether the difference is significant. The smaller the difference, the greater the likelihood that our hypothesized value for the mean is correct. The larger the difference, the smaller the likelihood.

The conventional approach to hypothesis testing is not to construct a single hypothesis about the population parameter, but rather to set up two different hypotheses. These hypotheses must be so constructed that if one hypothesis is accepted, the other is rejected and *vice versa*.

The two hypothesis in a statistical test are normally referred to as:

(i) Null hypothesis, and

(ii) Alternative hypothesis

The null hypothesis is a very useful tool in testing the significance of difference. In its simplest form the hypothesis asserts that there is no real difference in the sample and the population in the particular matter under consideration (hence the word "null" which means invalid, void, or amounting to nothing) and that the difference found is accidental and unimportant arising out of fluctuations of sampling. The null hypothesis is akin to the legal principle that a man is innocent until he is proved guilty. It constitutes a challenge; and the function of the experiment is to give the facts a chance to refute (or fail to refute) this challenge. For example, if we want to find out whether extra coaching has benefited the students or not, we shall set up a null hypothesis that "extra coaching has not benefited the students". Similarly, if we want to find out whether a particular drug is effective in curing malaria we will take the null hypothesis that "the drug is not effective in curing malaria". The rejection of the null hypothesis indicates that the differences have statistical significance and the acceptance of the null hypothesis indicates that the differences are due to chance. Since many practical problems aim at establishment of statistical significance of differences, rejection of the null hypothesis may thus indicate success in statistical project.

As against the null hypothesis, the alternative hypothesis specifies those values that the researcher believers to hold true, and of course, he hopes that the sample data lead to acceptance of this hypothesis as true. The alternative hypothesis may embrace the whole range of values rather than single point. Now a days, it is usually accepted common practice not to associate any special meaning to the null or alternative hypothesis but merely to let these terms represent to different assumptions about the population parameter. However, for statistical convenience it will make a difference as to which hypothesis is called the null hypothesis and which is called the alternative.

The null and alternative hypotheses are distinguished by the use of two different symbols, H_0 representing the null hypothesis and H_a the alternative hypothesis. Thus a psychologist who wishes to test whether or not a certain class of people have mean I.Q. higher than 100 might establish the following null and alternative hypotheses:

$H_0 : \mu = 100$ (null hypothesis)

$H_a : \mu \neq 100$ (alternative hypothesis).

Or, if he is interested in testing the differences between the mean I.Q. of two groups, this psychologist may like to establish the null hypothesis that the two groups have equal means ($\mu_1 - \mu_2 = 0$) and the alternative hypothesis that their means are not equal ($\mu_1 - \mu_2 \neq 0$)

$H_0 : \mu_1 - \mu_2 = 0$ (null hypothesis)

$H_a : \mu_1 - \mu_2 \neq 0$ (alternative hypothesis).

(2) *Set yo a Suitable Significance Level* : Having set up the hypothesis, the next step is to test the validity of H_0 against that of H_a at as certain level of significance. The confidence with which an experimenter rejects–or retains–a null hypothesis depends upon the significance level adopted. The significance level is customarily expressed as a percentage, such as 5 per cent, is the probability of rejecting the null hypothesis if it is true. When the hypothesis in question is accepted at the 5 per cent level, the statistician is running the risk that, in the long run, he will be making the wrong decision about 5 per cent of the time. By rejecting the hypothesis at the same level the runs the risk of rejecting a true hypothesis in 5 out of every 100 occasions. By testing at the 1 per cent level he seeks to reduce the chance of making a false judgment but some element of risk remains (1 out of 100 occasions) that he will make the wrong decision, *i.e.*, he may accept where he ought to have rejected or *vice-versa*.

The following diagram illustrates the region in which we would accept or reject the null hypothesis when it is being tested at 5 per cent level of significance and a two-tail test is employed.

The above diagram illustrates how to interpret a 5 per cent level of significance. It may be noted that 2.5 per cent of the area under the curve is located in each tail.

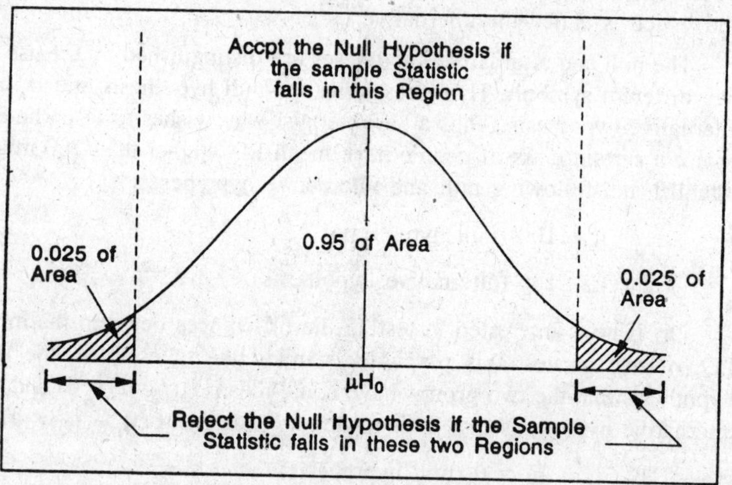

(3) *Setting a Test Criterion* : The third step in hypotheses testing procedure is to construct a test criterion. This involves selecting an appropriate probability distribution for the particular test, that is, a probability distribution which can properly be applied. Some probability distributions that are commonly used in testing procedures are t, F and c^2. Test criteria must employ an appropriate probability distribution; for example, if only small sample information is available, the use of the normal distribution would be inappropriate.

(4) *Doing Computations* : Having taken the first three steps, we have completely designed a statistical test. We now proceed to the fourth step–performance of various computations–from a random sample of size n, necessary for the test. These calculations include the testing statistic and the standard error of the testing statistic.

(5) *Making Decisions* : Finally, as a fifth step, we may draw statistical conclusions and take decisions. A statistical conclusion or statistical decision is a decision either to reject or to accept the null hypothesis. The decision will depend on whether the computed value of the test criterion falls in the region of rejection or the region of acceptance. If the hypothesis is being tested at 5 per cent level and the observed set of results has probabilities less than 5 per cent, we consider significant. In other words, we think that the sample result is so rare that it cannot be explained by chance variation alone. We then decide to reject H_0 and state: "the null hypothesis is false", or "the sample observations are not consistent

with the null hypothesis" (the rejection of H_0 automatically leads to acceptance of H_a).

On the other hand, if at 5 per cent level of significance the observed set of results has probability more than 5 per cent we give reason that the difference between the sample result and the hypothetical parameter can be explained by chance variations and, therefore, is not significant statistically. Consequently, we decide not to reject H_0 and state: "The sample observations are not inconsistent with the null hypothesis." If the probability is about 5 per cent, the wisest course may be to resolve judgment and draw another sample, if possible.

The reader might have noted above that the rejection statement is much stronger than the acceptance statement. In other words, if the null hypothesis is not rejected, the statistician does not then categorically conclude that the hypothesis is true. The difference in attitudes arises essentially from the fact that, in logic, it is always easier to prove something false than to prove it true.

It should be clearly noted that the practical "managerial decision" is outside the responsibility of the statistician. He does not make the decision; he purely provides information on the basis of which the businessman or administrator can be assisted in making his decisions

Two Types of Errors in Testing of Hypothesis

When a statistical hypothesis is tested there are four possibilities:

1. The hypothesis is true but our test rejects. (Type I error)
2. The hypothesis is false but our test accepts it. (Type II error)
3. The hypothesis is true but our test accepts it. (Correct decision)
4. The hypothesis is false but our test rejects it. (Correct decision)

Obviously, the first two possibility lead to errors.

In a statistical hypothesis testing experiment, a Type I error is committed by rejecting the null hypothesis when it is true. The probability of committing a Type I error is denoted by a α (pronounced as alpha), where

α = Prob. (Type I error)

= Prob. (Rejecting H_0/H_a is true)

On the other hand, a Type II error is committed by not rejecting (*i.e.*, accepting) the null hypothesis when it is false. The probability of committing a Type II error is denoted by β (pronounced as beta), where

β = Probability (Type II error)

= Probability (Not rejecting or accepting H_0/H_a is false)

The distinction between these two types of errors can be made by an example. Assume that the difference between two population mean is actually zero. If our test of significance when applied to the sample means leads us to believe that the difference in population means is significant, we make a Type I error. On the other hand, suppose there is true difference between the two population means. Now if our test of significance leads to the judgment "not significant", we commit a Type II error. We thus find ourselves in the situation which is described by the following table :

	Accept H_0	Reject H_a
H_0 is True	Correct Decision	Type I Error
H_0 is False	Type II Error	Correct Decision

While testing hypothesis the aim is to reduce both the types of error, *i.e.*, Type I and Type II. But due to fixed sample size, it is not possible to control both the errors simultaneously, Thère is a *trade-off* between these types of errors: the probability of making one type of error can only be reduced if we are willing to increase the probability of making the other type of error. In order to get a low β, we will have to put up with a high α. To deal with this trade-off in business situations, managers decide the appropriate level of significance by examining the costs or penalties attached to both types of errors.

It is more dangerous to accept a false hypothesis (Type II error than to reject a correct one (Type I error). Hence we keep the probability of committing Type I error at a certain level, called the level of significance. The level of significance (also known as the size of the rejection region or size of the critical region or simply size of the test) is traditionally denoted by the Greek letter α. In most statistical tests, the level of significance is generally fixed at 5 per cent. This means that the probability of accepting a true hypothesis is 95 per cent.

Two-tailed and One-tailed Tests of Hypothesis

While testing hypothesis we often talk of two-tailed tests and one-tailed tests. A two-tailed test of hypothesis will reject the null hypothesis, if the sample statistic is significantly higher than or lower than the hypothesized population parameter. Thus in a two-tail test the rejection region is located in both the tails. If we are testing a hypothesis at 5 per cent level of significance, the size of the acceptance region on each side of the mean would be 0.475 and the size of the rejection region is 0.025. If we consult the table of areas under the normal curve we find that an area of 0 475 corresponds to 1.96 standard errors on each side of μ_H, the hypothetical mean, and this equals the size of the acceptance region. If the sample mean falls into this area, the hypothesis is accepted. If the sample mean falls in the area beyond 1.96 standard error, the hypothesis is rejected because it falls into the rejection region. The acceptance and rejection regions for testing hypothesis, at the .05 level of significance, are given for a two-tail test.

It would be clear from the diagram that in a two-tail test, rejection regions are located in both tails.

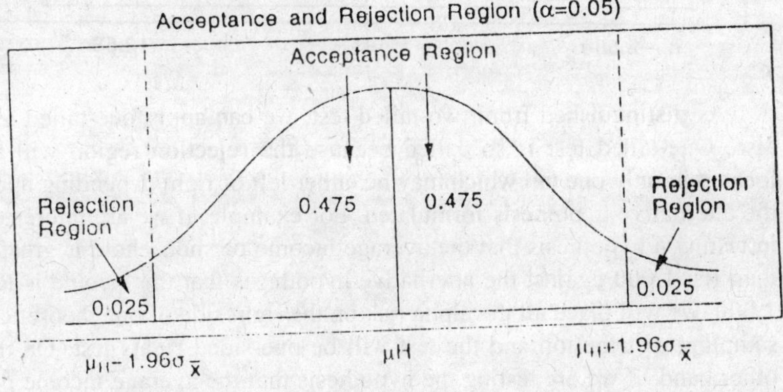

Suppose we want to reduce the risk of committing an error of type 1. This is done by reducing the size of the rejection region. For this a hypothesis may be treated at the .01 level of significance which means that the probability of rejecting a true hypothesis is 1 per cent. If we consult the table of areas under the normal curve. We find that an acceptance region of .495 (one half of .99) is equal to 2.58 standard error from μ_H. The acceptance and rejection regions at 0.01 level of significance are given in the diagram.

It will be clear from the above figure that as we decrease the size of rejection region, we increase the probability of accepting our hypothesis. As a = 0.01 the probability of rejecting true hypothesis is 1 per cent.

A two-tailed test is appropriate when the null hypothesis is $\mu = \mu_H$ (μ_H being some specified value) and the alternative hypothesis is $\mu = \mu_{H0}$. For example, if we are interested in testing the null hypothesis that the average income per household is Rs. 1,000 against the alternative hypothesis that it is not Rs., 1,000 the rejection region would lie on both sides because we would reject the null hypothesis if the mean income in the sample is either too far above Rs. 1,000 or to far below Rs. 1,000. Hence we are using a two-tail test.

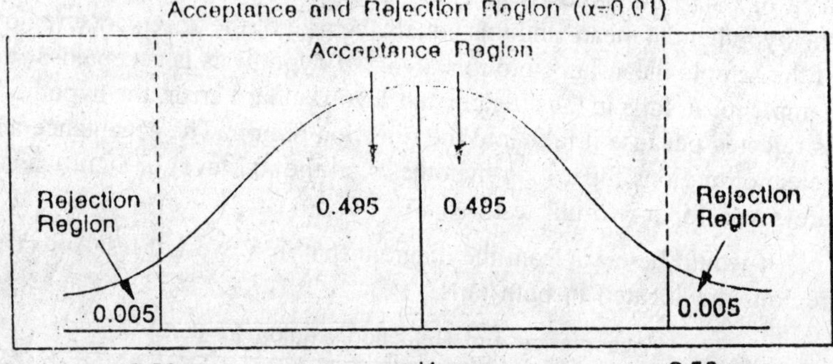

As distinguished from two-tailed test, we can apply one-tailed test also. One-tailed test is so called because the rejection region will be located in only one tail which may be either left or right depending upon the alternative hypothesis formulated. For example, if we are interested in testing a hypothesis that the average income per household is greater than Rs. 1,000 against the alternative hypothesis that the income is Rs. 1,000, we will place all the alpha risk on the right side of one theoretical sampling distribution and the test will be one-sided rights test. On the other hand, if we are testing the hypothesis that the average income per household is Rs., 1,000 against alternative that the income is less than Rs. 1,000 or less, the alpha risk is on the left side of theoretical sampling distribution and the test will be one-sided left tail test.

To sum up if we want to test the hypothesis that the population has a specified mean, say, μ_0 then the null hypothesis would be:

$$H_0 : \mu = \mu_0$$

and the alternative hypothesis could be:

(i) $H_1 : \mu \neq \mu_0$ (*i.e.*, $\mu > \mu_0$ or $\mu < \mu_0$)

(ii) $H_1 : \mu > m_0$

(iii) $H_1 : \mu < \mu_0$

The alternative hypothesis in (i) is known as a two-tailed alternative and the alternative in (ii) and (iii) are known as right-tailed and left-tailed alternatives. Accordingly, the corresponding tests of significance are called two-tailed, right-tailed and left-tailed tests respectively.

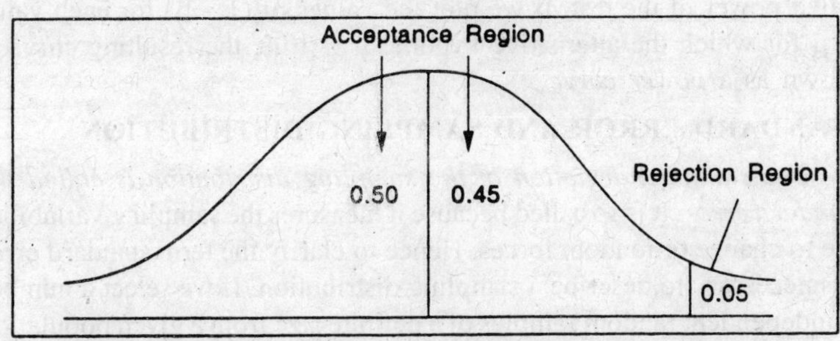

Acceptance And Rejection Region (One-Tail Test, Left α=0.05)

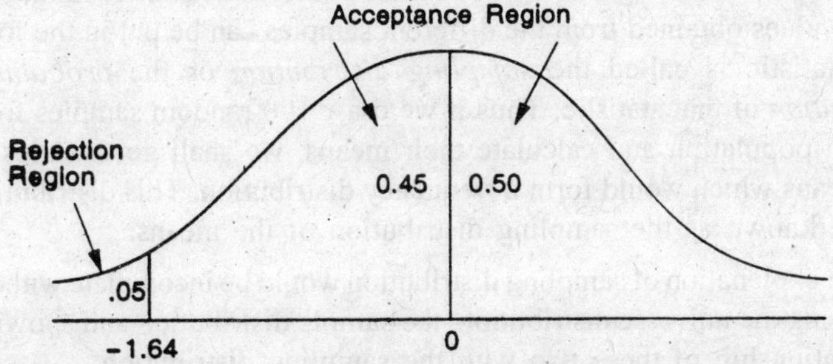

Measuring the Power of a Hypothesis Test

It is quite important to know how well a hypothesis test is working. The measure of how well the test is working, is called the *power of the test.*

In hypothesis testing α and β (the probabilities of type I and type II errors) should both be small. Type I error occurs when we reject a null hypothesis that is true and α (the significance level of the test) is the probability of making a type I error. Once a significance level is decided nothing can be done about α. Type II error occurs when we accept a null hypothesis that is false, the probability of type II error is β. The smaller the β, the better it is. Alternatively $(1 - \beta)$, *i.e.*, the probability of rejecting a null hypothesis when it is false, should be as large as possible.

Since rejecting a null hypothesis when it is false is exactly what a good test ought to do, a high value of $1 - \beta$ (something near 1) means the test is working quite well (it is rejecting the null hypothesis when it is false). A low value of $1 - \beta$ (something near 0) means that the test is working very poorly (it is not rejecting the null hypothesis when it is false). (1 – β) is the measure of how well the test is working and is called power of the test. If we plot the values of (1 – β) for each value of μ for which the alternative hypothesis is true, the resulting curve is known as a *power curve.*

STANDARD ERROR AND SAMPLING DISTRIBUTION

The standard deviation of the sampling distribution is called the standard error. It is so called because it measures the sampling variability due to chance or random forces. Hence to clarify the term standard error it is necessary to describe a sampling distribution. If we select a number of independent random samples of a definite size from a given population and calculate some statistic (lie the mean, standard deviation, etc.) from each sample, we shall get a series of values of these statistics or functions. These values obtained from the different samples can be put in the form of a statistic is called the *sampling distribution* or the *probability distribution* of that statistic. Thus if we draw 100 random samples from a given population and calculate their means, we shall get a series of 100 means which would form a frequency distribution. This distribution will be known as the sampling distribution of the means.

An explanation of sampling distribution would be incomplete without describing the universe distribution, the sample distribution and showing the relationship of these two with the sampling distribution.

Universe Distribution

Such a distribution emerges when each and every item of the universe is studied, and we have full knowledge of its mean and standard deviation. The mean of the universe which is also called the true mean is symbolized by μ (the lower-case m in Greek and called mu) and its standard deviation by σ (lower case sigma). Greek letters are used for these measures to emphasize their difference from corresponding measures taken from a sample. Measures characterizing a universe, such as μ and σ, are called parameters.

The Sample Distribution

If instead of all the items of the universe we study only a small part of it *i.e.*, take a sample, we will arrive at a sample distribution. The symbols $\overline{X}$ and S are used to designate the mean and standard deviation of the sample distribution. A measure characterizing a sample such as $\overline{X}$ or S, is called a statistic. It may be noted that several sample distributions are possible from a given universe.

The sampling distribution of a statistic reveals some important features:

(1) First, a sampling distribution is generated from a population distribution, known or assumed.

(2) Secondly, the same population may generate an infinite number of sampling distributions for the statistic, each for special sample size n.

(3) Finally, a population may generate distributions for two or more different statistics.

Sampling distributions are of great importance in theory and practice of statistics. It is because of the fact that the sampling distribution of a statistic has well-defined properties and it is from these properties that we can calculate risks (errors due to chance) involved in making generalisation about populations on the basis of samples.

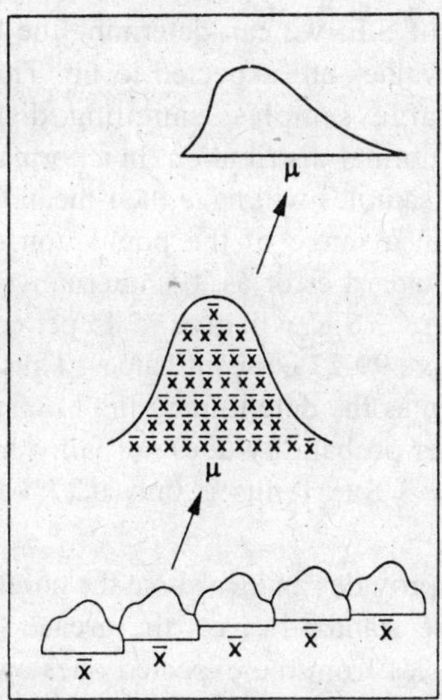

The above diagram would clearly illustrate the relationship between the three distributions :

For example, the sampling distributions of mean has the following important properties :

(1) The sampling distribution of means is normally distributed. Since the sampling distribution of the mean is normally distributed, we can develop a method in which we can use the mean of our sample to estimate the population mean. For example, if we compute an interval of $\overline{X} \pm 1.96\ \sigma\sqrt{n}$, the interval will include 95 per cent of all the sample means.

(2) The sampling distribution of mean has a standard deviation (a standard error) equal to the population standard deviation divided by the square root of the sample size, *i.e.*, $\frac{\sigma}{\sqrt{n}}$.

(3) The arithmetic mean of the sampling distribution is the same as the mean of the universe from which samples were taken. For this reason the mean of the sampling distribution may be denoted by the same symbol as that for the mean of the universe involved, namely μ.

Utility of Concept of Standard Error

The standard error is of great significance in statistical work due to following facts:

(1) With the help of S.E. we can determine the limits within which the parameter values are expected to lie. This is made possible because for large samples, sampling distributions tend to approximate a normal distribution. In a normal distribution 68.27 per cent of the samples will have their mean values (or any other constant) within a range of the population mean ± 1 standard deviation or standard error as it is alternatively called. Similarly a range of mean ± 2 S.E. will give 95.45 per cent values and mean ± 3 S.E. will give 99.73 per cent values. Thus a range of ± 3 S.E. should be taken as the determining limit outside which the value of the parameter probability does not fall. The chance of a value lying outside ± 3 S.E. limits is only 0.27% (*i.e.*, approximately 3 in 1,000).

(2) Standard error provides an idea about the unreliability of a sample. The greater the standard error, the greater is the departure of actual frequencies from the expected ones and hence the greater the unreliability of the sample. The reciprocal of S.E., *i.e.*, $\frac{1}{\text{S.E.}}$, is a measure of reliability (or precision) of the sample. The reliability or precision of an observed proportion varies as the square root of the number of items in the sample. In other words, if we want to double the precision (which is the same thing as reducing the standard error to one-half) the number of observations should be increased four times.

(3) It is used as an instrument in testing a given hypothesis. The hypotheses are generally tested at 5 per cent level of significance. If the difference between observed and expected means is more than 1.90 standard error (S.E.), we say that the result of the

experiment does not support the hypothesis at 5 per cent level or, in other words, the difference is regarded as significant, *i.e.*, it could not have arisen due to fluctuations of sampling. On the other hand, if the difference between observed and expected results is less than 1.96 S.E., it is not regarded as significant, *i.e.*, it could have arisen due to fluctuations of simple sampling, *i.e.*, we say that the result of the experiment does not provide any evidence against the hypothesis. If the difference is more than 2.58 S.E., it is considered to be significant at 1 per cent level. In practice, quite often a hypothesis is accepted if the difference is less than 3 S.E., because the probability of a difference greater than 3 S.E., arising by chance, is only about 3 in thousand (0.27%) as 99.73 per cent items are covered between mean ± 3s on either side of the mean. However, it must be emphasised that the use of 3s rule is justified only if n is large. Some people apply a criterion of 2 S.E. in order to determine whether or not the difference could have arisen due to fluctuations of sampling. However, instead of 3 S.E. or 2 S.E., it is suggested that we should use either 5 per cent level or 1 per cent level of significance. (In practice 5% level is more popular).

ESTIMATION

When data are collected by sampling from a population, the most important objective of statistical analysis is to draw inferences or generalisations about that population from the information embodied in the sample. Statistical estimation, or briefly estimation, is concerned with the methods by which population characteristics are estimated from sample information. It may be pointed out that the true value of a parameter is an unknown constant that can be correctly ascertained only by an exhaustive study of the population. However, it is ordinarily too expensive or it is infeasible to enumerate complete populations to obtain the required information. In case of finite populations, the cost of complete censuses may be prohibitive and in case of infinite population, complete enumerations are impossible. A realistic objective may be to obtain a guess or estimate of the unknown true value or an interval of plausible values from the sample data and also to determine the accuracy of the procedure. Statistical estimation procedures provide us with the means of obtaining estimates of population parameters with desired degrees of precision. With respect to estimating a parameter*. The following two types of estimates are possible:

1. Point estimates, and

2. Interval estimates.

1. Point Estimates

A point estimate is a single number which is used as an estimate of the unknown population parameter. The procedure in point estimation is to select a random sample of n observations, $x_1, x_2...x_3$ from a population $f(x, \theta^*)$ and then to use some preconceived method to arrive from these observations at a number say $\hat{\theta}$ (read theta hat) which we accept as an estimator of θ. The estimator θ is a single point on the real number scale and thus the name point estimation. $\hat{\theta}$ depends on the random variables that generate the sample and hence, it too is a random variable with its own sampling distribution.

2. Interval Estimates

As distinguished from a point estimate which provides one single value of the parameter, an interval estimate of a population parameter is a statement of two values between which it is estimated that the parameter lies. An interval estimate would always be specified by two values, *i.e.*, the lower one and the upper one. In more technical terms, interval estimation refers to the estimation of a parameter by a random interval, called the *Confidence interval,* whose end points L and U with $L < U$, are functions of the observed random variables such that the probability that the inequality $L < \theta < U$ is satisfied in terms of predetermined number, $1 - a$, L and U are called the confidence limits and are the random end points of interval estimate. Since in an interval estimate, we determine an interval of plausible values, hence the name interval estimation. Thus, on the basis of sample study, if we estimate the average income of the people living in a village as Rs. 875 it will be a point estimate. On the other hand, if we say that the average income could lie between Rs. 800 and Rs. 950, it will be an interval estimate.

PROPERTIES OF A GOOD ESTIMATOR

A distinction is made between an *estimate* and an *estimator*. The numerical value of the sample mean is said to be an estimate of the population mean figure. On the other hand, the statistical measure used, that is, the method of estimation, is referred to as an *estimator.* For example, the sample mean, $\overline{X}$ is an estimator of the population mean.

A good estimator, as common sense dictates, is close to the parameter being estimated. Its quality is to be evaluated in terms of the following properties :

(i) *Unbiasedness.* An estimator is said to be unbiased if its expected value is identical with population parameter being estimated. That is if $\hat{\theta}$ is an unbiased estimate of θ, then we must have E ($\hat{\theta}$) = θ. Many estimators are "Asymptotically unbiased" in the sense that the biases reduce to practically insignificant values zero when n becomes sufficiently large. The estimator S^2 is an example.

It should be noted that bias in estimation is not necessarily undesirable. It may turn out to be an asset in some situations. For example, it may happen that an unbiased estimator is less desirable than a biased estimator if the former has a greater variability than the latter and, as a consequence, the expected value of the latter is closer than that of the former to the parameter being estimated.

(ii) *Consistency.* If an estimator, say $\hat{\theta}$, approaches the parameter q closer and closer as the sample size n increases. $\hat{\theta}$ is said to be a consistent estimator of θ. Stating somewhat more regorously, the estimator $\hat{\theta}$ is said to be a consistent estimator of θ if, as n approaches infinity, the probability approaches I that $\hat{\theta}$ will differ from the parameter θ by not more than an arbitrary small constant.

The sample mean is an unbiased estimator of μ no matter what form the population distribution assumes, while the sample median is an unbiased estimate of m only if the population distribution is symmetrical. The sample mean is better than the sample median as an estimate of μ in terms of both unbiasedness and consistency.

In case of large samples consistency is a desirable property for an estimator to possess. However in small samples, consistency is of little importance unless the limit of probability defining consistency is reached even with a relatively small size of the simple.

(iii) *Efficiency.* The concept of efficiency refers to the sampling variability of an estimator. If two competing estimators are both unbiased, the one with the smaller variance (for a given sample size) is said to be relatively more efficient. Stated in a somewhat different language, estimator $\hat{\theta}_1$ is said to be more efficient than another estimator $\hat{\theta}_2$ for θ if the variance of the first is less than the variance of the second. The smaller the variance of the estimator, the more concentrated is the distribution of the estimator around the parameter being estimated and, therefore, the better this estimator is.

(iv) *Sufficiency.* An estimator is said to be sufficient if it conveys as much information as is possible about the parameter which is contained in the sample. The significance of sufficiency lies in the fact that if a sufficient estimator exists, it is absolutely unnecessary to consider a sample can furnish with respect to the estimation of a parameter is being utilised.

Many methods have been devised for estimating parameters that may provide estimators satisfying these properties. The two important methods are the least square method and the method of maximum likelihood.

Having discussed the above concepts let us now discuss the various situations where we have to apply different tests of significance. For the sake of convenience and clarity these situations may be summed up under the following three heads :

* Test of Significance for Attributes.
* Tests of Significance for Variables (Large Samples).
* Tests of Significance for Variables (Small Samples).

The various tests of significance for attributes are discussed under the following heads :

(1) Tests for Number of Successes,

(2) Tests for Proportion of Successes,

(3) Tests for Difference between Properties.

1. **Tests for Number of Successes.** The sampling distribution of the number of successes follows a binomial probability distribution. Hence its standard error is given by the formula :

S. E. of no. of successes = $\sqrt{npq}$

where n = size of sample

p = probability of success in each trial

q = (1 – p), *i.e.*, probability of failure.

Example 1:

(a) A coin was tossed 400 times and the head turned up 216 times. Test the hypothesis that the coin is unbiased.

Solution:

Let us take the hypothesis that the coin is unbiased. On the basis of this hypothesis the probability of getting head or tail would be equal, *i.e.*, 1/2. Hence in 400 throws of a coin we should expect 200 heads and 200 tails.

Observed number of heads = 216

Difference between observed number of heads and expected number of heads = 216 – 200 = 16

S.E. of no. of heads = $\sqrt{npq}$

$$n = 400,\ p = q = \frac{1}{2}$$

$$\therefore \text{S.E.} = \sqrt{400 \times \frac{1}{2} \times \frac{1}{2}} = 10$$

$$\frac{\text{Difference}}{\text{S.E.}} = \frac{16}{10} = 1.6$$

The difference between observed and expected number of heads is less than 1.96 S.E. (5% level of significance). Hence our hypothesis is true and we conclude that the coin is unbiased.

(b) In 324 throws of a xis-faced dice, odd points appeared 180 times. Would you say that the dice is fair at 5 per cent level of significance?

Solution:

Let us take the hypothesis that the dice is fair. In a fair dice we would expect 162 odd points in 324 throws

$$\text{S.E.} = \sqrt{npq} = \sqrt{324 \times \frac{1}{2} \times \frac{1}{2}} = 9$$

$$\frac{\text{Diff.}}{\text{S.E.}} = \frac{180-162}{9} = 2$$

Since the difference is more than 1.96 at 5 per cent level of significance, the hypothesis rejected. Hence we cannot say that the dice is fair at 5 per cent level of significance.

Example 2:

In a sample of 500 people from a village in Rajasthan, 280 are found to be rice eaters and the rest wheat eaters. Can we assume that both the food articles are equally popular?

Solution:

Let us take the hypothesis that the food articles are equally popular. The expected frequency for wheat eaters and rice eaters would be $\frac{500}{2} = 250$ each.

$$\text{S.E.} = \sqrt{npq} = \sqrt{500 \times \frac{1}{2} \times \frac{1}{2}} = 11.18$$

Difference between observed and expected number of rice eaters = 280 – 250 = 30

$$\frac{\text{Diff.}}{\text{S.E.}} = \frac{30}{11.18} = 2.68\,.$$

Since the difference is more than 2.58 S.E. (1% level), it could not have arisen due to fluctuations of sampling. Hence we cannot assume that both the food articles are equally popular.

Example 3:

A person throws 10 dice 500 times and obtains 2560 times, 4, 5, or 6. Can this be attributed to fluctuations of sampling?

Solution:

Let us take the hypothesis that there is no significant difference in the sample results and the expected results.

$$\text{S.E.} = \sqrt{npq}$$

$$n = 5000,\ p = \frac{1}{2},\ \{q = \frac{1}{2}$$

$$\text{S.E.} = \sqrt{500 \times \frac{1}{2} \times \frac{1}{2}} = 35.36$$

$$\frac{\text{Difference}}{\text{S.E.}} = \frac{2560 - 2500}{35.36} = 1.7$$

Since the difference is less than 1.96 S.E. at 5% level of significance the hypothesis holds true. Hence the difference could be attributed to fluctuations of sampling.

Example 4:

In a hospital 480 female and 520 male babies were born in a week. Do these figures confirm the hypothesis that males and females are born in equal number?

Solution:

Let us take the hypothesis that male and female babies are born in equal number, *i.e.*, $p = \frac{1}{2}$

$$S.E. = \sqrt{npq} = \sqrt{1000 \times \frac{1}{2} \times \frac{1}{2}} = 15.81$$

Difference between observed and expected number of female babies = 520 – 500 = 20

$$\frac{\text{Difference}}{\text{S.E.}} = \frac{20}{15.81} = 1.265.$$

Since the difference is less than 1.96 S.E. (5% level) it can be concluded that the male and female babies are born in equal number.

(2) **Test for Proportion of Successes :** Instead of recording the number of successes in each sample, we might record the proportion successes, that is $\frac{1}{n}$th of the number of successes in each sample. As this would amount to dividing all figures of the record by n, the mean proportion of successes must be p, and the standard deviation of the proportion of successes $\sqrt{\frac{pq}{n}}$. Thus we have the following formula:

$$S.E._p = \sqrt{\frac{pq}{n}}.$$

Example 5:

A wholesaler in apples claims that only 4% of the apples supplied by him are defective. A random sample of 600 apples contained 36 defective apples. Test the claim of the wholesaler.

Solution:

The wholesaler claims that only 4% of the apples are defective. Hence the 95% confidence limits are given by $\overline{X} \pm 1.96$ S.E.

$$S.E._p = \sqrt{\frac{pq}{n}} = \sqrt{\frac{.96 \times .04}{600}} = 0.008$$

$$\begin{aligned} \text{95\% confidence limits} &= \overline{X} \pm 1.96 \text{ S.E.} \\ &= .96 \pm 1.96\ (0.008) \\ &= .96 \pm 0.0157 \\ &= 0.9443 \text{ to } 0.9757. \end{aligned}$$

Out of 600 apples, the good apples may vary between .9943 × 600 = 566.50 and 0.9757 × 600 = 585.42 *i.e.*, 576 and 585.

The number of defectives is thus expected to lie between 15 and 33. Since the actual number is 36, the wholesaler's claim that only 4% of his apples are defective cannot be accepted.

(3) **Test for Difference between Properties :** If two samples are drawn from different populations, we may be interested in finding out whether the difference between the proportion of successes is significant or not. In such a case we take the hypothesis that the difference between p_1, *i.e.*, the proportion of successes in one sample, and p_2, *i.e.*, the proportion of successes in another sample, is due to fluctuations of random sampling.

The standard error of the difference between proportions is calculated by applying the following formula :

$$S.E.\ (p_1 - p_2) = \sqrt{pq\left(\frac{1}{n_1} + \frac{1}{n_2}\right)}$$

where p = the pooled estimate of the actual proportion in the population. The value of p is obtained as follows :

$$p = \frac{n_1 p_1 + n_2 p_2}{n_1 + n_2} \text{ or } p = \frac{x_1 + x_2}{n_1 + n_2}$$

where x_1 and x_2 stand for the number of occurrences in the two samples of sizes n_1 and n_2 respectively.

If $\frac{p_1 - p_2}{S.E.}$ is less than 1.96 S.E. (5% level of significance), the difference is regarded as due to random sampling variation, *i.e.*, as not significant.

The following illustrations shall explain the method.

Example 6:

If it costs a rupee to draw one number of a sample, how much would it cost in sampling from a universe with mean 100 and standard deviation 10 to take sufficient number as to ensure that the mean of a sample would in a 5 per cent probability be within 0.01 per cent of the true? Find the extra cost necessary to double this precision.

Solution:

$$\overline{X} = 100\%$$

Difference between universe mean and sample mean = 0.01 per cent

$$S.E.\ \overline{X} = \frac{\sigma}{\sqrt{n}}$$

For 95% confidence, difference between sample mean and population mean should be equal to 1.96 S.E., *i.e.*,

$$1.96 \times \frac{10}{\sqrt{n}}$$

$$\therefore \quad \frac{1.96 \times 10}{\sqrt{n}} = 0.01 \text{ or } \sqrt{n} \times 0.01 = 19.6$$

$$\sqrt{n} = \frac{19.6}{0.01} = 1960 \text{ or } n = (1960)^2 = 38,41,600$$

Double the precision would imply that the difference between sample mean and population mean should be = 0.005 only.

$$\frac{1.96 \times 10}{\sqrt{n}} = 0.005 \text{ or } \sqrt{n} = \frac{19.6}{0.005} = 3920$$

$\therefore n = (3920)^2 = 1,53,66,400$

$\therefore$ Extra cost = 1,53,66,400 – 38,41,600 = Rs. 1,15,24,800

Hence in order to double the precision the extra cost required is Rs. 1,15,24,800.

Example 7:

A sample of 100 tyres is taken from a lot. The mean life of tyres is round to be 39,350 kms. with a standard deviation of 3,260. Could the sample come from a population with mean life of 40,000 kms? Establish 99% confidence limits within which the mean life of tyres is expected to lie.

Solution:

Let us take the hypothesis that there is no significant difference between the sample mean and the hypothetical population mean.

$$\text{S.E.} \quad \overline{x} = \frac{\sigma}{\sqrt{n}} = \frac{3260}{\sqrt{100}} = 326$$

$$\frac{\text{Diff.}}{\text{S.E.}} = \frac{40,000 - 39,350}{326} = \frac{650}{326} = 1.994$$

Since the difference is less than 2.58 S.E. (at 1% level of significance) the hypothesis is accepted and the difference could have arisen due to fluctuations of sampling.

99% confidence limits $\overline{X} \pm 2.58$ (S.E.)

$$39350 + 2.58\ (326)$$

$$39350 \pm 841.08$$

$$38508.92 - 40191.08$$

Hence the mean life of tyres is expected to lie between 38,509 and 40191.

Example 8:

Find the regression coefficient of Bombay prices over Calcutta prices from the following data and also calculate its standard error:

	Bombay Rs.	***Calcutta Rs.***
Average price per quintal of wheat	*120*	*130*
Standard deviation	*4*	*5*
Coefficient of correlation	*= 0.6*	
n	*= 100*	

Solution:

Regression coefficient of Bombay prices (X) over Calcutta prices (Y)

$$b_{xy} = r\frac{\sigma_x}{\sigma_y} = 0.6\ \frac{4}{5} = 0.48$$

The standard error of regression coefficient is

$$S.E._{b_{xy}} = \frac{\sigma_x\sqrt{1-r^2}}{\sigma_y\sqrt{n}} = \frac{4\sqrt{1-(0.6)^2}}{5\sqrt{100}} = \frac{4 \times 0.8}{50} = 0.064$$

Thus the regression coefficient of Bombay prices over Calcutta prices is 0.48 and its standard error is 0.064.

Two-tailed test for Difference between the Means of Two Samples

(1) If two random samples with $\overline{X}_1, \sigma_1, n_1$ and $\overline{X}_2, \sigma_2, n_2$ respectively are drawn from different populations, then the S.E. of the difference between the mean is given by the formula :

$$= \sqrt{\frac{\sigma_1^2}{n_1} + \frac{\sigma_2^2}{n_2}}$$

and where σ_1 and σ_2 are unknown.

S.E. of the difference between means

$$= \sqrt{\frac{S_1^2}{n_1} + \frac{S_2^2}{n_2}}$$

where S_1 and S_2 represent standard deviations of the two samples.

The null hypothesis to be tested is that there is no significant difference in the means of the two samples, *i.e.*,

$H_0 : \mu_1 = \mu_2 \leftarrow$ null hypothesis, there is no difference

$H_1 : \mu_1 \neq \mu_2 \leftarrow$ alternative hypothesis, a difference exists.

Example 9:

An auto company decided to introduce a new six cylinder car whose mean petrol consumption is claimed to be lower than that of the existing auto engine. It was found that the mean petrol consumption for the 50 cars was 10 km per litre with a standard deviation of 3.5 km per litre. Test for the company at 5% level of significance, whether the claim the new car petrol consumption is 9.5 km per litre on the average is acceptable.

Solution:

Let us take the hypothesis that there is no significant difference between the sample average and the company's claim, *i.e.*,

$$H_0 = \overline{X} - m' = 0$$

$$\text{S.E. } \overline{X} = \frac{\sigma}{\sqrt{n}} = \frac{3.5}{\sqrt{50}} = \frac{3.5}{7.07} = 0.495$$

$$\frac{\text{Difference}}{\text{S.E.}\overline{X}} = \frac{10 - 9.5}{0.495} = 1.01$$

Since the difference is less than 1.96 S.E. at 5% level of significance, the hypothesis is accepted. Hence the company's claim that the new car petrol consumption is 9.5 km. per litre is acceptable.

Example 10:

Two samples of 100 electric bulbs each has a means 1500 and 1550, Standard deviation 50 and 60. Can it be concluded that two brands differ significantly at 1% level of significance in equality.

Solution:

Let us take the hypothesis that there is no significant difference in the mean life of the two makes of bulbs.

$$\text{S.E. } \left(\overline{X} - \overline{X}_2\right) = \sqrt{\frac{\sigma_1^2}{n_1} + \frac{\sigma_2^2}{n_2}}$$

$$= \sqrt{\frac{(50)^2}{100} + \frac{(60)^2}{100}} = \sqrt{25 + 36} = 7.81$$

$$\frac{\text{Difference}}{\text{S.E.}} = \frac{1550 - 1500}{7.81} = 6.4$$

Since the difference is more than 2.58 S.E. (1% level of significance), the hypothesis is rejected. Hence there is a significant difference in the mean life of the two brands of bulbs.

Example 11:

Intelligence test on two groups of boys and girls gave the following results :

	Mean	*S.D.*	*N*
Girls	*75*	*15*	*150*
Boys	*70*	*20*	*250*

Is there a significant difference in the mean scores obtained by boys and girls?

Solution:

Let us take the hypothesis that there is no significant difference in the mean scores obtained by boys and girls.

$$\text{S.E. } \left(\overline{X} - \overline{X}_2\right) = \sqrt{\frac{\sigma_1^2}{n_1} + \frac{\sigma_2^2}{n_2}}$$

$\sigma_1 = 15$, $\sigma_2 = 20$, $n_1 = 150$, $n_2 = 250$.

Substituting the values

$$\text{S.E., } \left(\overline{X} - \overline{X}_2\right) = \sqrt{\frac{(15)^2}{150} + \frac{(20)^2}{250}} = \sqrt{1.5 + 1.6} = 1.761$$

$$\frac{\text{Difference}}{\text{S.E.}} = \frac{75 - 70}{1.781} = 2.84$$

Since the difference is more than 2.58 S.E. (1% level of significance), the hypothesis is rejected. There seems to be a significant difference in the mean scores obtained by boys and girls.

Example 12:

The mean produce of wheat of a sample of 100 fields is 200 lb. per acre with a standard deviation of 10 lb. Another sample of 150 fields gives the mean at 220 with a standard deviation of lb. Assuming the standard deviation of the mean field at 1 lb. of the universe, find at 1% level if the two results are consistent.

Solution:

$$S.E._{(\sigma_1-\sigma_2)} = \sqrt{\frac{\sigma^2}{2}\left(\frac{1}{n_1}+\frac{1}{n_2}\right)} = \sqrt{\frac{(11)^2}{2}\left(\frac{1}{100}+\frac{1}{150}\right)}$$

$$= \sqrt{\frac{60.5}{100}+\frac{60.5}{150}} = \sqrt{0.605 + .403} = 1.004$$

Difference between two standard deviations = (12 – 10) = 2 lb.

$$\frac{\text{Difference}}{\text{S.E.}} = \frac{2}{1.004} = 1.992\,.$$

Since the difference is less than 2.58 S.E. (1% level of significance) the two results may be regarded as consistent.

$$S.E.\,\overline{x} = \frac{13.11}{\sqrt{100}} = \frac{13.11}{10} = 1.311\,.$$

Example 13:

The mean height obtained from a random sample of size 100 is 64 inches. The standard deviation of the distribution of height of the population is known to be 3 inches. Test the statement that the mean height of the population is 67 inches at 5% level of significance. Also set up 99% limits of the mean height of the population.

Solution:

Let us take the hypothesis that there is no significant difference between the sample mean and the population mean.

$$S.E.\,\overline{x} = \frac{\sigma}{\sqrt{n}} = \frac{3}{\sqrt{100}} = 0.3$$

$$\frac{\text{Diff.}}{\text{S.E.}} = \frac{67-64}{0.3} = 10$$

Since the difference is more than 1.96 S.E. (5% level) the hypothesis is rejected. Hence the mean height of the population could not be 67".

99% probable limits of the mean height of the population

$= \overline{X} \pm 2.58$ S.E.

$= 64 \pm 2.58\ (.3)$

$= 64 \pm 0.774 = 63.2$ to 64.8.

Example 14:

In a random sample of 1,000 persons from town A, 400 are found to be consumers of wheat. In a sample of 800 from town B, 400 are found to be consumers of wheat. Do these data reveal a significant difference between town A and town B, so far as the proportion of wheat consumers is concerned?

Solution:

Let us set up the hypothesis that the two towns do not differ so far as proportion of what consumers is concerned, *i.e.*, $H_0 : p_1 = p_2$ against $H_a : p_1 \neq p_2$.

Computing the standard error of the difference of proportions :

$$\text{S.E. } (p_1 - p_2) = \sqrt{pq\left(\frac{1}{n_1} + \frac{1}{n_2}\right)}$$

$$n_1 = 100,\ p_1 = \frac{400}{1000} = 0.4;\ n_2 = 800;\ p_2 = \frac{400}{800} = 0.5$$

$$p = \frac{(1000 \times 0.4) + (800 \times 0.5)}{1000 + 800} = \frac{400 + 400}{1800}$$

$$\text{or simply } p = \frac{x_1 + x_2}{n_1 + n_2} = \frac{400 + 400}{1000 + 800} = \frac{4}{9}$$

$$q = 1 - \frac{4}{9} = \frac{5}{9}$$

$$\text{S.E. } (p_1 - p_2) = \sqrt{\frac{4}{9} \times \frac{5}{9}\left(\frac{1}{1000} + \frac{1}{800}\right)} = \sqrt{\frac{20}{21} \times \frac{9}{4000}} = 0.024$$

$$p_1 - p_2 = 0.4 - 0.5 = -0.1^*$$

$$\frac{\text{Difference}}{\text{S.E.}} = \frac{0.1}{0.024} = 4.17$$

Since the difference is more than 2.58 S.E. (1% level of significance). It could not have arisen due to fluctuations of sampling. Hence the data reveal a significant difference between town A and town B so far as the proportion of wheat consumers is concerned.

Example 15:

A man buys 50 electric bulbs of 'Philips' and 50 electric bulbs of 'HMT'. He finds that 'Philips' bulbs give an average life of 15000 hours with a standard deviation of 60 hours and 'HMT' bulbs gave an average life of 1512 hours with a standard deviation of 80 hours. Is there a significant difference in the mean life of the two makes of bulbs?

Solution:

Let us set up the hypothesis that there is no significant difference in the mean life of the two makes of bulbs. Calculating standard error of difference of means.

$$\text{S.E.}\left(\overline{X}-\overline{X}_2\right)=\sqrt{\frac{\sigma_1^2}{n_1}+\frac{\sigma_2^2}{n_2}}$$

$\sigma_1 = 60$, $n_1 = 50$, $s_2 = 80$, $n_2 = 50$

$$\text{S.E.}\left(\overline{X}_1-\overline{X}_2\right)=\sqrt{\frac{(60)^2}{50}+\frac{(80)^2}{50}}=\sqrt{\frac{3600+6400}{50}}$$

$$=\sqrt{200}=14.14$$

Observed difference between the two means = 512 – 1500 = 12

$$\frac{\text{Difference}}{\text{S.E.}}=\frac{1}{14.14}=0.849$$

Since the difference is less than 2.58 S.E. (1% level of significance), it could have arisen due to fluctuations of sampling. Hence the difference in the mean life of the two makes is not significant.

In case of two large random samples, each drawn from a normally distributed population, the S.E. of the difference between the standard deviations is given by :

$$\text{S.E.}_{(\sigma_1-\sigma_2)}=\sqrt{\frac{\sigma_1^2}{2n_1}+\frac{\sigma_2^2}{2n_2}}$$

Where population standard deviations are not known

$$\text{S.E.}_{(S_1-S_2)}=\sqrt{\frac{S_1^2}{2n_1}+\frac{S_2^2}{2n_2}}$$

Example 16:

In a random sample of 1000 persons from U.P. 510 were found to be consumers of cigarettes. In another sample of 800 persons from Rajasthan, 480 were found to be consumers of cigarettes. Discuss the data reveal a significant difference between U.P. and Rajasthan so far as the proportion of consumers of cigarettes in concerned.

Solution:

Let us take the hypothesis that there is no significant difference between UP and Rajasthan so far as the proportion of consumers of cigarettes is concerned.

$$S.E.(p_1 - p_2) = \sqrt{pq\left(\frac{1}{n_1} + \frac{1}{n_2}\right)}$$

$$p_1 = \frac{510}{1000} = 0.51$$

$$p_2 = \frac{480}{800} = 0.60$$

$$p = \frac{x_1 + x_2}{n_1 + n_2} = \frac{510 + 480}{1000 + 800} - 0.55$$

$$q = 1 - p = 1 - .55 = .45$$

$$S.E. = \sqrt{.55 + .45\left(\frac{1}{1000} + \frac{1}{800}\right)} = 0.024$$

$$\frac{\text{Diff.}}{\text{S.E.}} = \frac{|.51 - .6|}{.024} = 3.75$$

Since the difference is more than 2.58 S.E. at 1% level of significance the hypothesis is rejected. Hence the data reveal a significant difference between UP and Rajasthan so far as proportion of consumers of cigarettes is concerned.

At times we may be interested in comparing the proportion of persons possessing an attribute in a sample with the proportion given by the population. In such a case the following formula is applicable:

$$S.E.\ (p_1 - p_2) = \sqrt{p_0 q_0 \times \frac{n_2}{n_1\,(n_1 + n_2)}}$$

where p_0 = Population proportion

$q_0 = 1 - p_0$

n_1 = Number of observations in the sample

$n_1 + n_2$ = Size of population

n_2 = (Size of population – n_1).

Example 17:

500 apples are taken at random from a large basket and 50 are found to be bad. Estimate the proportion of bad apples in the basket and assign limits within which the percentage most probably lies.

Solution:

The proportion of bad apples in the given sample $= \frac{50}{500} = 0.1$.

Hence p = 0.1 and q = 0.9

$$\text{S.E.} = \sqrt{\frac{pq}{n}} = \sqrt{\frac{0.1 \times 0.9}{500}} = \sqrt{\frac{0.09}{500}} = 0.013$$

The limits within which percentage of bad apples lies

$$= \left[p \pm 3\sqrt{\frac{pq}{n}}\right] \times 100$$

$= 0.1 \pm (0.013)] \times 100$

$= 0.1 \pm (0.039)] \times 100$

$= 6.1$ and 13.9

Thus the percentage of bad apples in the consignment almost certainly lies between 56.1 and 13.9.

Example 18:

In a simple random sample of 600 men taken from a big city 400 are found to be smokers. In another simple random sample of 900 men taken from another city 450 are smokers. Do the data indicate that there is a significant difference in the habit of smoking in the two cities?

(M. Com., Raj Univ., 1989; M. Com., Punjab Univ., 1996)

Solution:

Let us take the hypothesis that there is no significant difference in the habit of smoking in the two cities.

$p_1 - p_2 = 0.4 - 0.5 = -0.1$. However, the conclusion remains the same irrespective of the fact whether the difference is positive or negative.

$$\text{S.E. } (p_1 - p_2) = \sqrt{pq\left(\frac{1}{n_1} + \frac{1}{n_2}\right)}$$

$$p_1 = \frac{400}{600} = 0.667;\ p_2 = \frac{400}{900} = 0.5$$

$$p = \frac{x_1 + x_2}{n_1 + n_2},\ q = 1 - p$$

$$n_1 = 600,\ n_2 = 900,\ x_1 = 400,\ x_2 = 450$$

$$p = \frac{400 + 450}{600 + 900} = \frac{850}{1500} = \frac{17}{30}$$

$$q = 1 - \frac{17}{30} = \frac{13}{30}$$

$$\text{S.E. } (p_1 - p_2) = \sqrt{\frac{17}{30} \times \frac{13}{30}\left(\frac{1}{600} + \frac{1}{900}\right)}$$

$$= \sqrt{\frac{17}{30} \times \frac{13}{30} \times \frac{1500}{600 \times 900}} = 0.026$$

Since the difference is more than 2.58 S.E. at 1% level of significance, the hypothesis is rejected. Hence there is a significant difference in the habit of smoking of the two cities.

Example 19:

A machine produced 20 defective articles in a batch of 400. After overhauling it produced 10 defectives in a batch of 300. Has the machine improved?

Solution:

Let us take the hypothesis that the machine has not improved after overhauling. Applying test of difference of proportions.

$$p_1 = \frac{20}{400} = 0.050$$

$$p_2 = \frac{10}{300} = 0.033$$

Pooled estimate of actual proportion in the population.

$$p = \frac{x_1 + x_2}{n_1 + n_2} = \frac{20 + 10}{400 + 300} = 0.043$$

$$q = 1 - p = 1 - 0.43 = 0.957$$

$$\text{S.E. } (p_1 - p_2) = \sqrt{pq\left(\frac{1}{n_1} + \frac{1}{n_2}\right)}$$

$$= \sqrt{.043 \times .957\left(\frac{1}{400} + \frac{1}{300}\right)} = 0.0155$$

$$\frac{\text{Diff.}}{\text{S.E.}} = \frac{p_1 - p_2}{.0155} = \frac{0.05 - 0.033}{0.0155} = 1.1$$

Since the difference is less than 1.96 S.E. (at 5% level) our hypothesis is true. Machine has not improved after overhanding.

Example 20:

Before an increase in excise duty on tea 400 people of a sample of 500 persons were found to be tea drinkers. After an increase in the duty, 400 persons were known to be tea drinkers in a sample of 600 people. Do you think that there has been a significant decrease in the consumption of tea after the increase in the excise duty.

(M. Com., Delhi Univ., 1988; MBA, Delhi Univ. 1990, M. Com.; HPU, 1994)

Solution:

Let us take the hypothesis that there is no significant decrease in the consumption of tea after increase in the excise duty.

Computing the standard error of the difference of proportions

$$\text{S.E. } (P_1 - p_2) = \sqrt{pq\left(\frac{1}{n_1} + \frac{1}{n_2}\right)}$$

$$p_1 = \frac{400}{500} = 0.8;\ p_2 = \frac{400}{600} = 0.667$$

$$p = \frac{400 + 400}{500 + 600} = \frac{8}{11};\ q = 1 - \frac{8}{11} = \frac{3}{11}$$

$$\text{S.E. } (p_1 - p_2) = \sqrt{\frac{8}{11} \times \frac{3}{11}\left(\frac{1}{500} + \frac{1}{600}\right)} = 0.027$$

$$\frac{p_1 - p_2}{\text{S.E.}} = \frac{0.8 - 0.667}{0.027} = 4.93.$$

Since the difference is more than 2.58 S.E. (1% level of significance) it could not have arisen due to fluctuations of sampling. Hence, there is significant decrease in the consumption of tea after the increase in excise duty.

Example 21:

There are 1,000 students in a college. Out of 20,000 in the whole university, in a study 200 were found smokers in the college and 1,000 in the whole university. Is there a significant difference between the proportion of smokers in the college and university?

Solution:

Let us take the hypothesis that there is no significant difference in the proportion of smokers in the college and university. Proportion of smokers in the college.

Proportion of smokers in the university

$$= \frac{1000}{20000} = 0.05$$

Difference between the two proportions

$$= 0.2 - 0.05 = 0.15$$

p_0, *i.e.*, proportion of smokers in the university = 0.05

$q_0 = 1 - p_0 = 0.95$, $n_1 = 1000$,

$n_1 + n_2 = 20000$

So $n_2 = 20000 - 100 = 19000$

$$\text{S.E. of difference} = \sqrt{p_0 q_0 \times \frac{n_2}{n_1 (n_1 + n_2)}}$$

$$= \sqrt{0.05 \times 0.95 \left(\frac{19000}{1000 + 20000} \right)} = \sqrt{0.000015} = 0.0067$$

$$\frac{\text{Difference}}{\text{S.E.}} = \frac{0.15}{0.0067} = 22.39\,.$$

Since the difference is more than 2.58 S.E. (1% level of significance) it could not have arisen due to fluctuations of sampling. Hence there is a significant difference between the proportion of smokers in the college and university–the proportion being significantly higher in the college.

There are three main objects in studying problems relating to sampling of variables :

(1) To compare observation with expectation and to see how far the deviation of one from the other can be attributed to fluctuations of sampling;

(2) To estimate from samples some characteristic of the parent population, such as the mean of a variable; and

(3) To gauge the reliability of our estimates.

DIFFERENCE BETWEEN SMALL AND LARGE SAMPLES

It is normally agreed amongst statisticians that a sample is to be recorded as large only if its size exceeds 30. The tests of significance used for dealing with problems relating to large samples are different from the ones used for small samples for the reason that the assumptions that we make in case of large samples do not hold good for small samples. The assumptions made while dealing with problems relating to large samples are :

(i) The random sampling distribution of a statistic is approximately normal, and

(ii) Values given by the samples are sufficiently close to the population value and can be used in its place for calculating the standard error of the estimate.

While testing the significance of a statistic in case of large samples, the concept of standard error discussed earlier is used. The following is a list of the formulae for obtaining standard error for different statistics:

(1) Standard Error of Mean*

(i) When standard deviation of the population is known

$$S.E.\ \overline{X} = \frac{\sigma_p}{\sqrt{n}}$$

where S.E. $\overline{X}$ refers to the standard error of the mean

s_p = Standard deviation of the population

n = Number of observations in the sample.

(ii) When standard deviation of population is not known, we have to use standard deviation of the sample in calculating

* The standard error of the mean measures only sampling errors. Sampling errors are errors involved in estimating a population parameter from a sample instead of including all the essential information in the population.

standard error of mean. Consequently, the formula for calculating standard error is

$$\text{S.E. } \overline{X} = \frac{\sigma \text{ (sample)}}{\sqrt{n}}$$

where s denotes standard deviation of the sample.

It should be noted that if standard deviations of both sample as well as population are available then we should prefer standard deviation of the population for calculating standard error of mean.

Fiducial limits of population mean :

95% fiducial limits of population mean are

$$\overline{X} \pm 1.96 \frac{\sigma}{\sqrt{n}}$$

99% fiducial limits of population mean are

$$\overline{X} \pm 2.58 \frac{\sigma}{\sqrt{n}}$$

(2) S.E. of Median or $\text{S.E.}_{\text{Med}} = 1.25331 \frac{\sigma}{\sqrt{n}}$

(3) S.E. of Quartile or $\text{S.E.}_a = 1.36263 \frac{\sigma}{\sqrt{n}}$

(4) S.E. of Quartile Deviation of $\text{S.E.}_{\text{QD}} = 0.78672 \frac{\sigma}{\sqrt{n}}$

(5) S.E. of Mean Deviation or $\text{S.E.}_{\text{MD}} = 0.6028 \frac{\sigma}{\sqrt{n}}$

(6) S.E. of Standard Deviation or $\text{S.E.}_{\sigma} = \frac{\sigma}{\sqrt{2n}}$

(7) S.E. of Variance or $\text{S.E.}_{\sigma}{}^2 = \sigma^2 \sqrt{\frac{2}{n}}$

(8) S.E. of Coefficient of Correlation or $\text{S.E.}_r = \frac{1-r^2}{\sqrt{n}}$

(9) S.E. of Regression Coefficient of Y on X, *i.e.*, S.E. b_{yx}

$$= \frac{\sigma_x \sqrt{1-r^2}}{\sigma_x \sqrt{n}}$$

(10) S.E. of Regression Coefficient of X on Y, *i.e.*, S.E. b_{XY}

$$= \frac{\sigma_x \sqrt{1-r^2}}{\sigma_y \sqrt{n}}$$

(11) S.E. of Regression Estimate of Y on X or $S.E._{xy} = \sigma_x \sqrt{1-r^2}$

(12) S.E. of Regression Estimate of X on Y or $S.E._{yx} = \sigma_y \sqrt{1-r^2}$

(13) S.E. of Coefficient of Association, *i.e.*, S.E.Q.

$$= \frac{1-Q^2}{2}\sqrt{\frac{1}{(AB)} + \frac{1}{(A\beta)} + \frac{1}{(\alpha B)} + \frac{1}{(\alpha\beta)}}$$

(14) S.E. of Rank Correlation Coefficient, *i.e.*, $S.E._m$

$$= \frac{1}{\sqrt{n-1}}.$$

Example 22:

A simple sample of the height of 6,400 Englishmen has a mean of 67.85 inches and a standard deviation of 2.56 inches while a simple sample of heights of 1600 Austrians has a mean of 68.55 inches and standard deviation of 2.52 inches. Do the data indicate that the Austrians are on the average taller than the Englishmen? Give reasons for your answer.

Solution:

Let us take the hypothesis that there is no significant difference in the mean height of *Englishmen and Austrians.*

$$S.E.(\overline{X}_1 - \overline{X}_2) = \sqrt{\frac{\sigma_1^2}{n_1} + \frac{\sigma_2^2}{n_2}}$$

$$\sigma_1 = 2.56,\ n_1 = 6400,\ \sigma_2 = 2.52,\ n_2 = 1600$$

$$S.E. = \sqrt{\frac{(2.56)^2}{6400} + \frac{(2.52)^2}{1600}} = \sqrt{.001 + .004} = .0707$$

$$\frac{\text{Diff.}}{\text{S.E.}} = \frac{|67.85 - 68.55|}{.0707} = \frac{0.7}{.0707} = 9.9$$

Since the difference is more than S.E. (at 1% level of significance), the hypothesis is rejected. Hence the data indicates that the Austrians are on the average taller than the Englishmen.

Example 23:

(b) Strength test carried out on samples of two yarns spun to the same count gave the following results:

	Sample size	*Sample Mean*	*Sample Variance*
Yarn A	*4*	*52*	*42*
Yarn B	*9*	*42*	*56*

The strengths are expressed in pounds. Is the difference in mean strengths significant of the sources from which the samples are drawn?

Solution:

Let us take the hypothesis that the difference in the mean strength of the two yarns is not significant. Applying t-test of difference of means.

$$t = \frac{\overline{X}_1 - \overline{X}_2}{S} \sqrt{\frac{n_1 n_2}{n_1 + n_2}}$$

$$S = \sqrt{\frac{(n_1 - 1)\, S_1^2 + (n_2 - 1)\, S_2^2}{n_1 + n_2 - 2}}$$

$$= \sqrt{\frac{(4-1)\, 42 + (9-1)\, 56}{4+9-2}} = 7.22$$

$\overline{X}_1 = 52$, $\overline{X}_2 = 42$, $S = 7.22$, $n_1 = 4$, $n_2 = 9$

$$t = \frac{52 - 42}{7.22} \sqrt{\frac{4 \times 9}{4 + 9}}$$

$$= \frac{10}{7.22} \times 1.664 = 2.3$$

$v = 4 + 9 - 2 = 11$

For $v = 11$, $t_{0.05} = 1.796$

The calculated value of t is more than the table value. The hypothesis is rejected. Hence the difference in the mean strength of the two virus 'A' and 'B' is significant.

Example 24:

680 heads and 520 tails are obtained in tossing a coin 1,200 times. Can it be concluded that the coin is unbiased?

Solution:

Let us take the hypothesis that the coin is unbiased.

S.E. of number of successes $= \sqrt{npq}$

$n = 1{,}200, \; p = \frac{1}{2}, q = \frac{1}{2}$

$\therefore \text{ S.E.} = \sqrt{1{,}200 \times \frac{1}{2} \times \frac{1}{2}} = 17.32$

Difference between observed and expected number of heads

$= 680 - 600 = 80$

$\frac{\text{Diff.}}{\text{S.E.}} = \frac{80}{17.32} = 4.62$

Since the difference is more than 2.58 S.E. (1% level of significance), there is reason to doubt the hypothesis. Hence the coin seams to be biased.

Example 25:

Two samples of 6 and 5 items respectively gave the following data:

Mean of 1st sample	*40*
Standard deviation of 1st sample	*8*
Mean of the second sample	*50*
Standard deviation of the 2nd sample	*10*

Is the difference of means significant? The value of t for 9 degrees of freedom at 5% level is 2.26.

Solution:

Let us take the hypothesis that there is no significant difference in the mean of two samples. Applying t-test :

$$t = \frac{\overline{X}_1 - \overline{X}_2}{S} \sqrt{\frac{n_1 n_2}{n_1 + n_2}}$$

$\overline{X}_1 = 40, \; \overline{X}_2 = 50, \; n_1 = 6, \; n_2 = 5$

$$S = \sqrt{\frac{(n_1 - 1)\, S_1^2 + (n_2 - 1)\, S_2^2}{n_1 + n_2 - 2}}$$

$$= \sqrt{\frac{5(8)^2 + 4(10)^2}{6+5-2}} = \sqrt{\frac{320+400}{9}} = 8.944$$

$$t = \frac{40-50}{8.944}\sqrt{\frac{6 \times 5}{6+5}} = \frac{10}{8.944} \times 1.651 = 1.846$$

$v = 6 + 5 - 2 = 9$

For $v = 9$ $t_{0.05} = 2.26$

The calculated value of t is less than the table value. Hence the hypothesis holds true. We therefore, conclude that the difference of means of two samples is not significant.

Example 26:

10 persons were appointed in a electrical position in an office. Their performance was noted by giving a test and the means recorded out of 50. They were given 6 month's training and again they were given a test and marks were recorded out of 50.

Employees :	*A*	*B*	*C*	*D*	*E*	*F*	*G*	*H*	*I*	*J*
Before Training :	*25*	*20*	*35*	*15*	*42*	*28*	*26*	*44*	*35*	*48*
After Training :	*26*	*20*	*34*	*13*	*43*	*40*	*29*	*41*	*36*	*46*

By applying the t-test can it be concluded that employees have benefited by the training? (You are given for $v = 9$, $t_{0.05} = 2.262$*).*

Solution:

Let us take the hypothesis that the employees have not benefited by the training. Applying t-test :

Before training	After training	(2nd-1st)	d^2
25	26	+1	1
20	20	0	0
35	34	–1	1
15	13	–2	4
42	43	+1	1
28	40	+12	144
26	29	+3	9
44	41	–3	9
35	36	+1	1
48	46	–2	4
		$\Sigma d = 10$	$\Sigma d^2 = 174$

$$t = \frac{\bar{d}\sqrt{n}}{S}; \quad \bar{d} = \frac{\Sigma d}{n} = \frac{10}{10} = 1$$

$$S = \sqrt{\frac{\Sigma d^2 - n(\bar{d})^2}{n-1}} = \sqrt{\frac{174 - 10(1)^2}{9}} = 0.741$$

$$t = \frac{1 \times \sqrt{20}}{4.269} = \frac{3.1622}{4.269} = 0.741$$

$v = n - 1 = 10 - 1 = 9$

For $v = 9$, $t_{0.05} = 2.262$.

The calculated value of t is less than the table value and hence the hypothesis is accepted. The employees have not benefited from the training.

Example 27:

A simple random sample of size 100 has mean 15, the population variance being 25. Find an interval estimate of the population mean with a confidence level of (i) 99%, and (ii) 95%.

If population variance is not given, then what should be done to find out the required interval estimates?

Solution:

We are given N = 100

$\bar{X}$ = 15 and variance = 25

Since variance = 25, $\sigma = \sqrt{25} = 5$

$$S.E.\bar{x} = \frac{\sigma}{\sqrt{N}} = \frac{5}{\sqrt{100}} = \frac{5}{10} = 0.5$$

99% Confidence Limits

$\bar{X} \pm 2.58$ S.E.

15 ± 1.96 (.5) = 15 ± 1.29 or 13.71 to 16.29

95% Confidence Limits

$\bar{X} \pm 1.96$ S.E.

15 ± 1.96 (.5) = 15 ± 0.98 or 14.02 to 15.98

Example 28:

In a random sample of 500 persons belonging to urban area 200 are found to be commuters of public transport. In another sample of 400 persons belonging to rural area 200 are found to be commuters of public transport. Discuss whether the data repeat a significant difference between urban area to far as the proportion of commuters of public transport is concerned.

Solution:

Let us take the hypothesis that there is no significant difference between rural and urban area so far as the proportion of commuters of public transport is concerned. Applying test of difference of proportions.

$$S.E._{(p_1-p_2)} = \sqrt{pq\left(\frac{1}{n_1}+\frac{1}{n_2}\right)}$$

$$p_1 = \frac{200}{500} = 0.4$$

$$p_2 = \frac{200}{400} = 0.5$$

$$p = \frac{x_1+x_2}{n_1+n_2} = \frac{200+200}{500+400} = 0.444$$

$$q = 1 - p = 0.556$$

$$S.E._{(p_1-p_2)} = \sqrt{.444 \times .556\left(\frac{1}{500}+\frac{1}{400}\right)} = 0.033$$

$$\frac{\text{Diff.}}{\text{S.E.}} = \frac{|.4-.5|}{.033} = 3.03$$

Since the difference is more than 2.58 S.E. at 1% level of significance, the hypothesis is rejected. Hence there is significant difference in the rural and urban areas so far as the proportion of commuters of public transport is concerned.

Example 29:

A group of seven week-old chickens reared on a high protein diet weigh 12, 15, 11, 16, 14, 14 and 16 ounces, a second group of five chickens similarly treated except that they receive a low protein diet weighted 8, 10, 14, 10 and 13 ounces. Test whether there is sufficient evidence that additional protein has increased the weight of the chickens.

Solution:

Let us take the hypothesis that the additional protein has not increased the weight of chickens. Applying t-test :

$$t = \frac{\overline{X}_1 - \overline{X}_2}{S}\sqrt{\frac{n_1 n_2}{n_1 + n_2}}$$

X_1	$(X_1 - \overline{X}_1)$	$(X_1 - \overline{X}_1)^2$	X_2	$(X_2 - \overline{X}_2)$	$(X_2 - \overline{X}_2)^2$
12	–2	4	8	–3	9
15	+1	1	10	–1	1
11	–3	9	14	+3	9
16	+2	4	10	–1	1
14	0	0	13	+2	4
14	0	0			
16	+2	4			
$\Sigma X_1 = 98$	$\Sigma(X_1 - \overline{X}_1) = 0$	$\Sigma(X_1 - \overline{X}_1)^2 = 22$	$\Sigma X_2 = 55$	$\Sigma(X_2 - \overline{X}_2) = 0$	$\Sigma(X_2 - \overline{X}_2)^2 = 24$

$$\overline{X}_1 = \frac{98}{7} = 14;\ \overline{X}_2 = \frac{55}{5} = 11$$

$$S = \sqrt{\frac{\Sigma(X_1 - \overline{X}_1)^2 + \Sigma(X_2 - \overline{X}_2)^2}{n_1 + n_2 - 2}} = \sqrt{\frac{22+24}{7+5-2}} = \sqrt{\frac{46}{10}} = 2.14$$

$$t = \frac{14-11}{2.14}\sqrt{\frac{7 \times 5}{7+5}} = \frac{3}{2.14} \times 1.708 = 2.394$$

For $v = 10$, $t_{0.05} = 2.33$.

The calculated value is more than the table value. The hypothesis rejected. Hence there is significant evidence that additional protein has increased the weight of the chickens.

Example 30:

Sim-Trim, an agency conducting a weight reduction programme, claims that participants in their programme achieve a weight reduction of at least 5 kg. after two weeks of the programme.

In evidence thereof they have given the following data on 10 participants who had undergone this programme, about their weights in kg prior to the programme and two weeks after the programme.

On the basis of sample evidence, can the claim of the agency on weight reduction be substained?

Test at a significance level of 5%.

Before (kg.) 86 92 100 93 88 80 88 92 95 106
After (kg.) 77 84 92 87 80 74 80 85 95 96

Solution:

Let us take the hypothesis that there is no significant difference in weight reduction before and after the programme. Applying t-test:

$$t = \frac{\bar{d}\sqrt{n}}{S}$$

Before	After	d	d²
86	77	–9	81
92	84	–8	64
100	92	–8	64
93	87	–6	36
88	80	–8	64
80	74	–6	36
88	80	–8	64
92	85	–7	49
95	95	0	0
106	96	–10	100
		$\Sigma d = -70$	$\Sigma d^2 = 558$

$$\bar{d} = \frac{\Sigma d}{n} = \frac{70}{10} = -7$$

$$S = \sqrt{\frac{\Sigma d^2 - n(\bar{d})^2}{n-1}} = \sqrt{\frac{558 - 10\left(-7\right)^2}{9}} = 2.777$$

$v = n - 1 = 10 - 1 = 9$

For $v = 9$, $t_{0.05} = 2.28$.

The calculated value of t is higher than the table value. The hypothesis is rejected. Hence there is a significant in the weight before and after the programme.

Example 31:

Two salesman A and B are working in a certain district. From a sample survey conducted by the head office, the following results were obtained. State whether there is any significant difference in the average sales between the two salesmen.

	A	*B*
No. of sales	*10*	*18*
Average sales (in Rs.)	*170*	*205*
Standard Deviation (in Rs.)	*20*	*25*

Solution:

Let us take the hypothesis that there is no significant difference in the average sales of the two salesmen.

Applying t-test :

$$t = \frac{\overline{X}_1 - \overline{X}_2}{S}\sqrt{\frac{n_1 n_2}{n_1 + n_2}}$$

$\overline{X}_1 = 170$, $\overline{X}_2 = 205$, $n_1 = 10$, $n_2 = 18$, $S_1 = 20$, $S_2 = 25$.

$$S = \sqrt{\frac{(n_1 - 1)\,S_1^2 + (n_2 - 1)\,S_2^2}{n_1 + n_2 - 2}} = \sqrt{\frac{(10-1)\,400 + (18-1)625}{10+8-2}}$$

$$= \sqrt{\frac{3600 + 10625}{26}} = \sqrt{547.115} = 23.39$$

$$t = \frac{170 - 205}{23.39}\sqrt{\frac{10 \times 18}{10+18}} = 3.79$$

$u = n_1 + n_2 - 2 = 10 + 18 - 2 = 26.$

For $u = 26$, $t_{0.05} = 2.056$.

The calculated value of t is greater than the table value. The hypothesis is rejected. We, therefore, conclude that there is significant difference in the average sales of the two salesmen.

Example 32:

An investigation of the relative merits of two kinds to flashlight batteries showed that a random sample of 100 batteries of brand X lasted on the average 36.5 hours with a standard deviation of 1.8 hours, while a random sample of 80 batteries of brand Y, lasted on the average 36.8 hours with a standard deviation of 1.5 hours. Use a level of significance of 0.05 to test whether the observed difference between the average life times is significant.

Solution:

Let us take the hypothesis that the observed difference in the average life of the two makes of batteries does not differ significantly

$$S.E._{(\bar{x}_1-\bar{x}_2)} = \sqrt{\frac{\sigma_1^2}{n_1}+\frac{\sigma_2^2}{n_2}}$$

$\sigma_1 = 1.8,\ n_1 = 100,\ \sigma_2 = 1.5,\ n_2 = 80$

$$\therefore\ S.E._{(\bar{x}_1-\bar{x}_2)} = \sqrt{\frac{(1.8)^2}{100}+\frac{(1.5)^2}{80}} = 0.246$$

$$\frac{\text{Difference}}{\text{S.E.}} = \frac{36.5-36.81}{0.2343} = \frac{1.31}{0.246} = 1.22\,.$$

Since the difference is less than 1.96 S.E. (5% level) the hypothesis holds true. Hence the average life of the two makes of batteries does not (differ significantly).

Example 33:

To study the correlation between the stature of father and the stature of son, a sample of 1,600 is taken from the universe of fathers and sone. The sample study gives the correlation between the two to be 0.80. Within what limits does it hold true for the universe?

Solution:

The standard error of the correlation coefficient between the stature of the father and son is

$$S.E._{r} = \frac{1-r^2}{\sqrt{n}}$$

$r = 0.8,\ n = 1,600$

$$S.E._{r} = \frac{1-(0.8)^2}{\sqrt{1600}} = \frac{1-0.64}{40} = \frac{0.36}{40} = 0.009\,.$$

If the sampling was simple random sampling the correlation in the universe cannot deviate from the correlation in the sample by more than thrice the standard error, *i.e.*, .027. Hence correlation in the universe most probably lies between r ± 3 S.E. or S ± 0.027 or 0.773 and 0.827.

Example 34:

From the following data compute the value of Yule's Coefficient of Association and find out its standard error. Also determine 95 per cent fiducial limits of the true values in the population :

Married and failures = *100*

Married and non-failures = *50*

Unmarried and failures = *20*

Unmarried and non-failures = *80*

Solution :

Let A denote married and B failures

$\therefore$ α will denote unmarried and β non-failures. The given information in terms of these symbols is

$$(AB) = 100,\ (A\beta) = 50,\ (\alpha\beta) = 20,\ (\alpha\beta) = 80$$

$$Q = \frac{(AB)(\alpha\beta) - (A\beta)(\alpha B)}{(AB)(\alpha\beta) + (A\beta)(\alpha B)}$$

$$= \frac{(100 \times 80) - (50 \times 20)}{(100 \times 80) + (50 \times 20)} = \frac{7,000}{9,000} = +\ 0.778$$

$$\text{S.E.} = \frac{1-Q^2}{2}\sqrt{\frac{1}{(AB)} + \frac{1}{(A\beta)} + \frac{1}{(\alpha B)} + \frac{1}{(\alpha\beta)}}$$

$$= \left\{\frac{1-(0.778)^2}{2}\right\}\sqrt{\frac{1}{100} + \frac{1}{50} + \frac{1}{20} + \frac{1}{80}}$$

$$= \left\{\frac{1-0.605}{2}\right\}\sqrt{0.01 + 0.02 + 0.05 + 0.0125}$$

$$= 0.1975\sqrt{0.0925}$$

$$= 0.1975 \times 0.304 = 0.06.$$

95% fiducial limits :

Q ± 1.96 S.E. will give the required limits

0.778 ± 1.96 (0.06)

0.778 ± 0.1176 = 0.66 and 0.896.

Example 35:

The mean life time of a sample of 400 fluorescent light bulbs produced by a company is found to be 1,570 hours with a standard deviation of 150 hours. To the hypothesis that the mean life time of the bulbs produced by the company is 1600 hours against the alternative hypothesis that it is greater than 1,600 hours at 1% level of significance.

Solution:

Let us take the hypothesis that there is no significant difference between the sample mean and hypothetical population mean *i.e.* $X_1 - \mu = 0$

$$\text{S.E.}\bar{x} = \frac{\sigma}{\sqrt{n}} = \frac{150}{\sqrt{400}} = \frac{150}{20} = 7.5$$

$$\frac{\text{Diff.}}{\text{S.E.}} = \frac{[1570 - 1600]}{7.5} = 4$$

The difference is more than 2.58 S.E. at 1% level of significance. The hypothesis is rejected. Hence there is significant difference between sample mean and hypothetical population mean.

Example 36:

In a survey buying habits, 400 women shoppers are chosen at random in super market A located in a certain section of Bombay city. Their average monthly food expenditure is Rs. 250 with a standard deviation of Rs. 40. For 400 women shoppers chosen at random in super market B in another section of the city, the average monthly food expenditure is Rs. 220 with a standard deviation of Rs. 55. Test at 1% level of significance whether the average food expenditure of the two populations of shoppers from which the samples were obtained are equal.

Solution:

Let us take the hypothesis that there is no difference in the average food expenditure of the two populations of shoppers.

$$\text{S.E.}\left(\bar{X}_1 - \bar{X}_2\right) = \sqrt{\frac{\sigma_1^2}{n_1} + \frac{\sigma_2^2}{n_2}}$$

$$n = 400,\ \bar{X}_1 = 250,\ \sigma_1 = 40;\ n_1 = 400,\ \bar{X}_2 = 220,\ \sigma_2 = 55$$

Substituting the values

$$\text{S.E.}\left(\bar{X}_1 - \bar{X}_2\right) = \sqrt{\frac{(40)^2}{400} + \frac{(55)^2}{400}} = \sqrt{\frac{1600}{400} + \frac{3025}{400}} = 3.4$$

$$\text{Difference of Means} = \left(\bar{X}_1 - \bar{X}_2\right) = 250 - 220 = 30$$

$$\frac{\text{Difference}}{\text{S.E.}} = \frac{30}{3.4} = 8.82\,.$$

Since the difference is more than 2.58 S.E. (1% level) the hypothesis is rejected. Hence the average food expenditures of the populations of shoppers are not equal.

Example 37:

For a given group of adults, the coefficient of correlation between height and weight is 0.7, standard deviations of height and weight are 2 inches and 10 1b. respectively and the means of height and weight for the entire group are 70 inches and 130 lb. respectively. Find out the best estimate of the weight of an individual who is 65 inches tall. Assign limits to this estimate in which in all probability his actual weight would be lying.

Solution:

The regression of weight (Y) on height (X) is

$$Y - \quad Y - \overline{Y} = r\frac{\sigma_y}{\sigma_x}\left(X - \overline{X}\right)$$

$$\overline{Y} = 130\ \overline{X} = 70\,,\ s_y = 10,\ s_x = 2,\ r = 0.7$$

$$Y - 130 = 0.7\ \frac{10}{2}(X - 70)$$

$$Y - 130 = 3.5X - 245 \text{ or}$$

$$Y = 3.5X - 115$$

when $\qquad X = 65$, Y will be

$$Y = 3.5\ (65) - 115 = 112.5 \text{ lb.}$$

The standard error of estimate of Y on X or S.E._{yx} is :

$$\text{S.E.}_{yx} = \sigma_y\sqrt{1 - r^2}$$

$$= 10\sqrt{1-(0.7)^2} = 10\sqrt{0.51} = 10 \times 0.714 = 7.14 \text{ lb.}$$

Thus the best estimate of the weight of an individual who is 65 inches tall is 112.5 lb. and this estimate cannot deviate from the actual weight by more than three times the standard error. Hence in all probability his actual weight would be lying between 112.5 ± 35 (1.428) or 112.5 ± 4.284, *i.e.*, 108.2 lbs and 116.8 lbs.

Interested not so much in the value of the correlation in the parent population, but more generally whether this value could have arisen from an uncorrelated population, *i.e.* whether it is significant of correlation in the parent population.

THE ASSUMPTION OF NORMALITY

While dealing with small samples also, an assumption is made that the parent population is normal, unless otherwise stated. Strictly speaking, therefore, our results will be true only for the normal population. However, as pointed out earlier, the assumption of normality is not very much warranted in case of small samples. Experiments have, therefore, been made to ascertain whether the results are true for other types of population. Theoretical work confirms that the results remain true for populations which do not deviate markedly from normality. However, if there is any good reason to suspect that the parent population is markedly skew, *i.e.*, U, or J-shaped, the methods given below cannot be applied with much confidence.

Since in many of the problems it becomes necessary to take a small size sample, considerable attention has been paid in developing suitable tests for dealing with problems of small samples. The greatest contribution to the theory of small samples is that of Sir William Gosset and R.A. Fasher. Sir William Gosset published his discovery in 1905 under the pen name 'Student'. He gave a test popularly known as 't-test' and Fisher gave another test known as 'z-test'. These tests are based on 't'-distribution and ‘z’-distribution.

Student’s t-Distribution

Theoretical work on t-distribution was done by W.S. Gosset (1876 – 1937) in the early 1900. Gosset was employed by the Guinness & Son, a Dublin bravery, Ireland, which did not permit employees to publish research findings under their own names. So Gosset adopted the pen name "student" and published his findings under this name. Thereafter, the t-distribution is commonly called *Student's t-distribution* or simply *Student's distribution.*

The t-distribution is used when sample size is 30 or less and the population standard deviation is unknown.

The “t-statistic” is defined as :

$$t = \frac{\overline{X} - \mu}{S} \times \sqrt{n}$$

where $S = \sqrt{\dfrac{\Sigma\left(X-\overline{X}\right)^2}{n-1}}$

The t-distribution has been derived mathematically under the assumption of a normally distributed population.

It has the following form

$$f(t) = C\left(1+\frac{t^2}{v}\right)^{-(v+1)/2}$$

$$\text{where } t = \frac{\left(\overline{X}-\mu\right)}{S}\sqrt{n}$$

C = a constant required to make the area under the curve equal to unity v = n – 1, the number of degrees of freedom.

Properties of t-Distribution

(1) The variance of the t-distribution is greater than one, but approaches one as the number of degrees of freedom and, therefore, the sample size becomes large. Thus the variance of the t-distribution approaches the variance of the standard normal distribution as the sample size increases. It can be demonstrated that from an infinite number of degrees of freedom ($v - \infty$), the t-distribution and normal distribution are exactly equal. Hence there is a widely practised rule of thumb that samples of size $n > 30$ may be considered be used as an approximation to t-distribution, where the latter is the theoretically correct functional form.

(2) Like the standard normal distribution, the t-distribution is symmetrical and has a mean zero.

(3) The constant c is actually a function of v (pronounced as nu), so that for a particular value of v, the distribution of f(t) is completely specified. Thus f(t) is a family of functions, one for each value of v.

(4) The variable t-distribution ranges from minus infinity to plus infinity.

The following diagram compares one normal distribution with two t-distributions of different sample sizes :

The above diagram shows two important characteristics of t-distribution. First, a t-distribution is lower at the mean and higher at the tails than a normal distribution. Second, the t-distribution has proportionately greater area in its tails than the normal distribution. Interval widths from t-distributions are, therefore, wider than those based on the normal distribution.

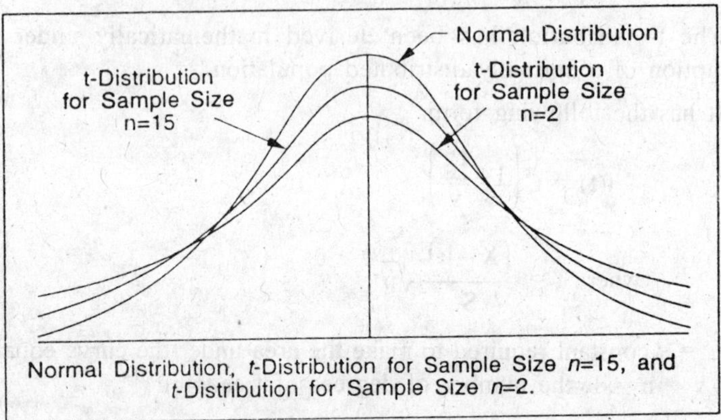

Normal Distribution, *t*-Distribution for Sample Size *n*=15, and *t*-Distribution for Sample Size *n*=2.

The t-Table

The t-table given at the end is the probability integral of t-distribution. It gives, over a range of values of v, the probabilities of exceeding by chance value of t at different levels of significance. The t-distribution has a different value for each degree of freedom and when degrees of freedom are infinitely large, the t-distribution is equivalent to normal distribution and the probabilities shown in the normal distribution tables are applicable.

Application of the t-Distribution

The following are some of the examples to illustrate the way in which the 'Student' distribution is generally used to test the significance of the various results obtained from small samples.

(1) *To Test the Significance of the Mean of a Random Sample:* In determining whether the mean of a sample drawn from a normal population deviates significantly from a stated value (the hypothetical value of the populations mean), when variance of the population is unknown we calculate the statistic :

$$t = \frac{(\overline{X} - \mu)\sqrt{n}}{S}$$

where $\overline{X}$ = the mean of the sample

μ = the actual or hypothetical mean of the population

n = the sample size

S = the standard deviation of the sample

$$X = \sqrt{\frac{\Sigma\left(X - \overline{X}\right)^2}{n-1}} \text{ or } S = \sqrt{\frac{\Sigma d^2 - n\left(\overline{d}\right)^2}{n-1}}$$

$$= \sqrt{\frac{1}{(n-1)}\left[\Sigma d^2 - \frac{(\Sigma d)^2}{n}\right]}$$

where d = deviation from the assumed mean.

If the calculated value of [t] exceeds $t_{0.05,}$ we say that the difference between $\overline{X}$ and m is significant at 5% level, if it exceeds $t_{0.01,}$ the difference is said to be significant at 1% level. If $|t| < t_{0.05,}$ we conclude that the difference between $\overline{X}$ and m is not significant and hence the sample might have been drawn from a population with mean = μ.

Fiducial Limits of Population Mean. Assuming that the sample is a random sample from a normal population of unknown mean the 95% fiducial limits of the population mean (μ) are :

$$\overline{X} \pm \frac{S}{\sqrt{n}} t_{0.05}$$

and 99% limits are

$$\overline{X} \pm \frac{S}{\sqrt{n}} t_{0.05}$$

The following examples will illustrate this test :

Example 38:

The life time of electric bulbs for a random sample of 10 from a large consignment gave the following data :

Item :	*1*	*2*	*3*	*4*	*5*	*6*	*7*	*8*	*9*	*10*
Life in '000 hours:	*4.2*	*4.6*	*4.1*	*5.2*	*3.8*	*3.9*	*4.3*	*4.4*	*5.6*	

Can we accept the hypothesis that the average life time of bulbs is 4,000 hours.

Solution:

Let us take the hypothesis that there is no significant difference in the sample mean and the hypothetical population mean. Applying the t-test.

$$t = \frac{\overline{X} - \mu}{S} \sqrt{n}$$

Calculation of $\overline{X}$ and S

X	$(X - \overline{X})$	$(X - \overline{X})^2$
4.2	–.02	0.04
4.6	+ 0.2	0.04
3.9	– 0.5	0.25
4.1	– 0.3	0.09
5.2	+ 0.8	0.64
3.8	– 0.6	0.36
3.9	– 0.5	0.25
4.3	– 0.1	0.01
4.4	0	0
5.6	+ 1.2	1.44
ΣX = 44		**$\Sigma (X - \overline{X})^2 = 3.12$**

$$\Sigma = \sqrt{\frac{\Sigma\left(X - \overline{X}\right)^2}{n-1}} = \sqrt{\frac{3.12}{9}} = 0.589$$

$$t = \frac{4.4 - 4}{0.589} \sqrt{10} = \frac{0.4 \times 3.162}{0.589} = 2.148$$

$v = n - 1 = 10 - 1 = 9$. For $v = 9$, $t_{0.05} = 2.262$

The calculated value of t is less than the table value. The hypothesis is accepted. The average life time of the bulbs could be 4000 hours.

Example 39:

Prices of shares of a company on different days in a month were found to be :

66, 65, 70, 69, 71, 70, 63, 64 and 68. Discuss whether the price of the shares be 65 (the table value of t = 2.262).

Solution:

Let us take the hypothesis that there is no difference between the mean price of the shares and the hypothetical mean. Applying t-test,

$$t = \frac{\overline{X} - \mu}{S} \sqrt{n}$$

Calculating Mean and Standard Deviation

X	(X – 67) d	d^2
66	–1	1
65	–2	4
69	+2	4
70	+3	9
69	+2	4
71	+4	16
70	+3	9
63	–4	16
64	–3	9
68	+1	1
ΣX = 675	**Σd = 5**	**Σd^2 = 73**

$$\overline{X} = \frac{675}{10} = 67.5; \ \overline{d} = \frac{5}{10} = 0.5$$

$$S = \sqrt{\frac{\Sigma d^2 - n(\overline{d})^2}{n-1}} = \sqrt{\frac{73-10\,(.5)^2}{9}} = 2.799$$

$$= \frac{67.5 - 65}{2.799} \times \sqrt{10} = 2.82$$

$v = 10 - 1 = 9$. For $v = 9$, $t_{0.05} = 2.262$.

The calculated value of t is greater than the table value. Our hypothesis does not hold good. Hence the mean price of the shares could not be Rs. 65.

Example 40:

Two laboratories A and B carry out independent estimates of fat content in ice-cream made by a firm. A sample is taken from each, halved, and the separated halves sent to the two laboratories. That fat content obtained by the laboratories is recorded below :

Bath No.	*1*	*2*	*3*	*4*	*5*	*6*	*7*	*8*	*9*	*10*
Lab. A	*7*	*8*	*7*	*3*	*8*	*6*	*9*	*4*	*7*	*8*
Lab. B	*9*	*8*	*8*	*4*	*7*	*7*	*9*	*6*	*6*	*6*

(He fact contents are given in grams.)

Is there a significant difference between the mean fat content obtained by the two laboratories A and B?

You may use the following extracts from t table in answering the question:

Degrees of freedom	*6*	*7*	*8*	*9*	*10*	*16*	*18*	*20*
5% value of t	*1.45*	*2.36*	*2.31*	*2.26*	*2.23*	*2.12*	*2.10*	*2.09*

Solution :

Let us take the hypothesis that there is no significant difference between the mean fat content obtained by the two laboratories, A and B, Applying t-test:

$$t = \frac{\overline{X}_1 - \overline{X}_2}{S} \times \sqrt{\frac{n_1 n_2}{n_1 + n_2}}$$

Since the actual means are in fractions, we shall take deviations from the assumed means.

	Lab. A			Lab. B		
Batch No.	X_1	$(X_1 - A_1)$ $A_1 = 16.$	$(X_1 - A_1)^2$	X_2	$(X_2 - A_2)$ $A_2 \neq 7$	$(X_2 - A_2)^2$
1	7	+1	1	9	+2	4
2	8	+2	4	8	+1	1
3	7	+1	1	8	+1	1
4	3	−3	9	4	−3	9
5	8	+2	4	7	0	0
6	6	0	0	7	0	0
7	9	+3	9	9	+2	4
8	4	−2	4	6	−1	1
9	7	+1	1	6	−1	1
10	8	+2	4	6	−1	1
	$\Sigma X_1 = 67$	$\Sigma(X_1 - A_1) = 7$	$\Sigma(X_1 - A_1)^2 = 37$	$\Sigma X_2 = 70$	$\Sigma(X_2 - A_2) = 0$	$\Sigma(X_2 - A_2)^2 = 22$

$$X_1 = \frac{67}{10} = 6.7; \overline{X}_2 = 7.0$$

$$S = \sqrt{\frac{\Sigma(X_1 - A_1)^2 + \Sigma(X_2 - A_2)^2 - n_1(\overline{X}_1 - A_1)^2 - n_2(\overline{X}_2 - A_2)^2}{n_1 + n_2 - 2}}$$

$$= \sqrt{\frac{37 + 22 - 10(6.7 - 6)^2 - 10(7 - 7)^2}{10 + 10 - 2}} = \sqrt{\frac{59.49}{18}} = \sqrt{\frac{54.1}{18}} = 1.734$$

$$t = \frac{6.7 - 7}{1.734}\sqrt{\frac{10 \times 10}{10 + 10}} = \frac{0.3}{1.734} \times 2.236 = 0.87$$

$v = n_1 + n_2 - 2 = 18$; For $v = 18$, $t_{0.05} = 2.10$.

The calculated value of |t| is less than the table value and hence there is reason to doubt the hypothesis. We, therefore, conclude that the mean fat contents obtained by two laboratories A and B do not differ significantly.

(2) *Testing Difference Between Means of Two Samples (Independent Samples)* : Given two independent random samples of size n_1 and n_2 with means $\overline{X}_1$ and $\overline{X}_2$ and standard deviations S_1 and S_2 we may be interested in testing the hypothesis that the samples come from the same normal population. To carry out the test, we calculate the statistic as follows :

$$t = \frac{\overline{X}_1 - \overline{X}_2}{S} \times \sqrt{\frac{n_1 n_2}{n_1 + n_2}}$$

where $\overline{X}_1$ = mean of the first sample

$\overline{X}_2$ = mean of the second sample

n_1 = number of observations in the first sample

n_2 = number of observations in the second sample

S = combined standard deviation.

The value of S is calculated by the following formula :

$$S = \sqrt{\frac{\Sigma(X_1 - \overline{X}_1)^2 + \Sigma(X_2 - \overline{X}_2)^2}{n_1 + n_2 - 2}}$$

When the actual means are in fraction the deviations should be taken from assumed means. In such a case the combined standard deviation is obtained by applying the following formula :

$$S = \sqrt{\frac{S(X_1 - A_1)^2 + S(X_2 - A_2)^2 - n_1(\overline{X} - A_1)^2 - n_2(\overline{X}_2 - A_2)^2}{n_1 + n_2 - 2}}$$

A_1 = Assumed mean of the first sample

A_2 = Assumed mean of the second sample

$\overline{X}_1$ = Actual mean of the first sample

$\overline{X}_2$ = Actual mean of the second sample.

The degrees of freedom = $(n_1 + n_2 - 2)$

When we are given the number of observations and standard deviation of the two samples, the pooled estimate of standard deviation can be obtained can be obtained as follows :

$$S = \sqrt{\frac{(n_1 - 1)\, S_1^2 + (n_2 - 1)\, S_2^2}{n_1 + n_2 - 2}}$$

If the calculated value of t be > $t_{0.05}$ ($t_{0.01}$), the difference between the sample means is said to be significant at 5% (1%) level of significance otherwise the data are said to be consistent with the hypothesis.

Example 41:

In a sample of 1000 the mean is 17.5 and the s.d. 2.5. In another sample of 800 the mean is 18 and s.d. 2.7. Assuming that the samples are independent discuss whether the two samples can have come from a population which have the same s.d.

Solution :

Let us take the hypothesis that there is no significant difference in the standard deviations of the two samples

$$S.E._{(\sigma_1 - \sigma_2)} = \sqrt{\frac{\sigma_1^2}{2n_1} + \frac{\sigma_2^2}{2n_2}}$$

$\sigma_1 = 2.5$, $n_1 = 1000$, $\sigma_2 = 2.7$, $n_2 = 800$

$$S.E._{(\sigma_1 - \sigma_2)} = \sqrt{\frac{(2.5)^2}{2000} + \frac{(2.7)^2}{1600}} = \sqrt{\frac{6.25}{2000} + \frac{7.29}{1600}}$$

$$= \sqrt{0.00312 + 0.00456} = 0.0876$$

$$\frac{\text{Diff.}}{\text{S.E.}} = \frac{2.7 - 2.5}{0.0876} = \frac{.2}{.0876} = 2.283$$

Since the difference is more than 1.96 S.E. at 5% level of significance the hypothesis rejected. Hence the two samples have not come from a population which has the same standard deviation.

Example 42:

To verify whether a course in accounting improved performance, a similar test was given to 12 participants both before and after the course. The original marks recorded in alphabetical order of the participants – were 44, 40, 61, 52, 44, 70, 41, 67, 72, 53, and 72. After the course, the marks were in the same order, 53, 38, 69, 57, 46, 39, 73, 48, 73, 74, 60 and 78. Was the course useful?

Solution:

Let us take the hypothesis that there is no difference in the marks obtained before and after the course, *i.e.*, the course has not been useful.

Applying t-test (difference formula) :

$$t = \frac{\bar{d}\sqrt{n}}{S}$$

Participants	Before (1st Test)	After (2nd Test)	(2nd–1st Test d	d²
A	44	53	+9	81
B	40	38	–2	4
C	61	69	+8	64
D	52	57	+5	25
E	32	46	+14	196
F	44	39	–5	25
G	70	73	+3	9
H	41	48	+7	49
I	67	73	+6	36
J	72	74	+2	4
K	53	60	+7	49
L	72	78	+6	36
			$\Sigma d = 60$	$\Sigma d^2 = 578$

$$\bar{d} = \frac{\Sigma d}{n} = \frac{60}{12} = 5$$

$$S = \sqrt{\frac{\Sigma d^2 - n(\bar{d})^2}{n-1}} = \sqrt{\frac{578 - 12(5)^2}{12-1}} = \sqrt{\frac{278}{11}} = 5.03$$

$$t = \frac{5 \times \sqrt{12}}{5.03} = \frac{5 \times 3.464}{5.03} = 3.443$$

$v = n - 1 = 12 - 1 = 11$; For $v = 11$, $t_{0.05} = 2.201$

The calculated value of t is greater than the table value. The hypothesis is rejected. Hence the course has been useful.

(3) *Testing Difference between Means of Two Samples (Dependent Samples or Matched Paired Observations)* : In the previous test it was assumed that the two samples were independent *i.e.*, the values of observations in one sample does not depend on the other. However, there are many situations in which this condition does not hold true, *i.e.*, we have dependent (or paired) samples. Two samples are said to be dependent when the elements in one sample are related to those in the other in any significant or meaningful manner. In fact, the two samples may consist of pairs of observations made on the same object, individual or, more generally, on the same selected population elements. When samples are dependent they comprise the same number of elementary units. We may carry out some experiment, say, to find out the effect of training on some employees, find out the efficacy of a coaching class or determine whether there is a significant difference in the efficacy of two drugs–one made within the country and another imported. Often the use of dependent or paired samples will enable us to perform a more precise analysis, because they will allow us to control the entraneous factors. The t-test based on paired observations is defined by the following formula :

$$t = \frac{\bar{d} - 0}{S} \times \sqrt{n} \text{ or } t = \frac{\bar{d}\sqrt{n}}{S}$$

where $\bar{d}$ = the mean of the differences

S = the standard deviation of the differences

The value of S is calculated as follows :

$$S = \sqrt{\frac{\Sigma(d - \bar{d})^2}{n-1}} \text{ or } \sqrt{\frac{\Sigma d^2 - n(\bar{d})^2}{n-1}}$$

It should be noted that t is based on n – 1 degrees of freedom.

The following examples will illustrate the application of difference test:

Example 43:

The manufacturer of a certain make of electric bulbs claims that his bulbs have a mean life of 25 months with a standard deviation of 5 months A random sample of 6 such bulbs gave the following values.

Life of months 24, 26, 30, 20, 20 18.

Can you regard the producer's claim to be valid at 1% level of significance? (Given that the table values of the appropriate test statistic at the said level are 4.032, 3.707 and 3.499 for 5, 6 and 7 degrees of freedom respectively).

Solution:

Let us take the hypothesis that there is no significant difference in the mean life of bulbs in the sample and that of the population. Applying t-test :

$$t = \frac{(\overline{X} - \mu)}{S}\sqrt{n}$$

Calculation of $\overline{X}$ and S

X	$(X - \overline{X})$ x	x^2
24	+1	1
26	+3	9
30	+7	49
20	–3	9
20	–3	9
18	–5	25
ΣX = 138		$\Sigma x^2 = 102$

$$\overline{X} = \frac{\Sigma X}{n} = \frac{138}{6} = 23$$

$$S = \sqrt{\frac{\Sigma x^2}{n-1}} = \sqrt{\frac{102}{5}} = \sqrt{20.4} = 4.517$$

$$= \frac{23 - 25}{4.517}\sqrt{6} = \frac{2 \times 2.449}{4.517} = 1.084$$

$v = n - 1 = 6 - 1 = 5$. For $v = 5$, $t_{0.01} = 4.032$.

The calculated value of t is less than the table value. The hypothesis is accepted. Hence the producer's claim is not valid at 1% level of significance.

Example 44:

Is a correlation coefficient of 0.5 significant if obtained from a random sample of 11 pairs of values from a normal population? Use t-test.

Solution:

Let the null hypothesis be that there is no correlation or the given correlation coefficient is not significant.

$$t = \frac{r}{\sqrt{1-r^2}} \times \sqrt{n-2}$$

$r = 0.5$, $n = 11$

$$t = \frac{0.5}{\sqrt{1-(0.5)^2}} \times \sqrt{11-2} = \frac{0.5}{\sqrt{1-0.25}} \times = \frac{0.5}{0.866} \times 3 = 1.732$$

$v = 11 - 2 = 9$ For $v = 9$, $t_{0.05} = 2.26$

The calculated value of t is less than the table value and hence the given correlation coefficient is not significant.

Example 45:

Two working designs are under consideration for adoption in a plant. A time and motion study shows that 12 workers using design A have a mean assembly time of 300 seconds with a standard deviation of 12 seconds and that 15 workers using design B have a mean assembly time of 335 seconds with a standard deviation of 15 seconds. Is the difference in the mean assembly time between the two working designs significant at one per cent level of significance? The following table gives some t-values which may be used:

Level of significance	*Degrees of Freedom*		
	25	*26*	*27*
$P_r = 0.05$	*2.06*	*2.06*	*2.05*
$P_r = 0.01$	*2.79*	*2.78*	*2.77*

Solution:

Let us take the hypothesis that the difference between the two designs 'A' and 'B' in respect of mean assembly time is not significant.

Applying t-test:

$$t = \frac{\overline{X}_1 - \overline{X}_2}{S} \sqrt{\frac{n_1 n_2}{n_1 + n_2}}$$

$$\overline{X}_1 = 300, S_1 = 12, n_1 = 12$$

$$\overline{X}_2 = 335, S_2 = 15, n_2 = 15$$

$$S = \sqrt{\frac{(n_1 - 1) S_1^2 + (n - 2) S_2^2}{n_1 + n_2 - 2}}$$

$$S = \sqrt{\frac{11(12)^2 + 14(15)^2}{12 + 15 - 2}} = \sqrt{\frac{5.103}{25}} = 13.8$$

$$t = \frac{|300 - 335|}{13.8} \times \sqrt{\frac{12 \times 15}{12 + 15}} = \frac{35}{13.8} \times 2.582 = 6.55$$

$$v = n_1 + n_2 - 2 = 12 + 15 - 2 = 25$$

For $v = 25, t_{0.01} = 2.79$

The calculated value of v is greater than the table value. Hence the hypothesis does not hold true. We, therefore, conclude that the difference in the mean assembly time between the two working designs is significant at 1 per cent level.

Cautions while using t-test : While drawing inference on the basis of t-test it should be remembered that the conclusions arrived at on the basis of the 't-test' are justified only if the assumptions upon which the test is based are true. If the actual distribution is not normally distributed then, strictly speaking the t-test is not justified for small samples. If it is not a random sample, then the assumption that the observations are statistically independent is not justified and the conclusions based on the t-test may not be correct. The effect of violating the normality assumption is slight when making inference about means provided that the sampling is fairly large when dealing with small samples. However, it is a good idea to check the normality assumption, if possible. A review of similar samples or related research may provide evidence as to whether or not the population is normally distributed.

Z-test of the Significance of the Correlation Coefficient : Prof. Fisher has given a method of testing the significance of the correlation coefficient in small samples. According to this method the coefficient of correlation is transformed into Z and hence the name Z-*transformation.* The statistic Z given by Prof. Fisher is used to test (i) whether an observed value of r differs significantly from some hypothetical value, or (i) whether two sample values of r differ significantly.

For testing whether r differs significantly from zero, the t-test is preferable.

In order to apply the test we have to calculate Z and ξ by applying Fisher's transformation and then calculate the value of the standard normal variate.

$$\frac{Z-\xi}{1/\sqrt{n-3}}$$

If the absolute value of this statistic exceeds 1.96, the difference is significant at 5% level.

Here $$Z = \frac{1}{2}\log_e \frac{1+r}{1-r} \text{ or } 1.1513 \log_{10}\left(\frac{1+r}{1-r}\right)$$

$$\xi = \frac{1}{2}\log_e \frac{1+\rho}{1-\rho} \text{ or } 1.1513 \log_{10}\left(\frac{1+\rho}{1-\rho}\right)$$

r refers to the population correlation coefficient.

$$S.E._{z} = \frac{1}{\sqrt{n-3}}.$$

Example 46:

12 students were given intensive coaching and 5 tests were conducted in a month. The scores of tests 1 and 5 are given below. Does the scores from the 1 to 5 show an improvement?

No. of Students	*1*	*2*	*3*	*4*	*5*	*6*	*7*	*8*	*9*	*10*	*11*
Marks in 1st test	50	42	51	26	35	42	60	41	70	55	62
Marks in 5th test	62	40	61	35	30	52	68	51	84	63	72

(The value of 'r' for 11 degrees of freedom at 5% level of significance is 2.20).

Solution:

Let us take the hypothesis that there is no difference in the scores obtained in the first and fifth test. Applying t-test :

$$t = \frac{\bar{d}\sqrt{n}}{S}$$

No. of Students	Marks in 1st Test	Marks in 5th Test	(5th – 1st Test) d	d^2
1	50	62	+12	144
2	42	40	–2	4
3	51	61	+10	100
4	26	35	+9	81
5	35	30	–5	25
6	42	52	+10	100
7	60	68	+8	64
8	41	51	+10	100
9	70	84	+14	196
10	55	63	+8	64
11	62	72	+10	100
12	38	50	+12	144
			$\Sigma d = 96$	$\Sigma d^2 = 1122$

$$\bar{d} = \frac{\Sigma d}{n} = \frac{96}{12} = 8$$

$$S = \sqrt{\frac{\Sigma d^2 - n(\bar{d})^2}{n-1}}$$

$$= \sqrt{\frac{1122 - 12(8)^2}{12-1}}$$

$$= 5.673$$

$$t = \frac{8\sqrt{12}}{5.673} = 4.885$$

For $v = 11,\ t_{0.05} = 2.20.$

The calculated value of t is greater than the table value hypothesis stands rejected. Hence the scores from test 1 to 5 show an improvement.

Example 47:

A random sample of 27 pairs of observations from a normal population gives a correlation coefficient of 0.42. Is it likely that the variables in the population are uncorrelated?

Solution:

Let us take the hypothesis that there is no significant difference in the sample correlation and correlation in the population. Applying t-test:

$$t = \frac{r}{\sqrt{1-r^2}} \times \sqrt{n-2}$$

Here $r = 0.42, n = 27$

$$t = \frac{.42}{\sqrt{1-(.42)^2}} \times \sqrt{27-2}$$

$$= \frac{.42}{.908} \times 5 = 2.31$$

$$v = n - 2 = 27 - 2 = 25$$

For $v = 25$, $t_{0.05} = 1.708$. The calculated value of t is more than the table value. The hypothesis is rejected. Hence it is likely that the variables in the population are uncorrelated.

Example 48:

For a random sample of 10 persons, fed on diet A, the increased weight in pounds in certain period were:

10, 6, 16, 17, 13, 12, 8, 14, 15, 9,

For another random sample of 12 persons, fed on diet B, the increase in the same period were :

7, 13, 22, 15, 12, 14, 18, 8, 21, 23, 10, 17

Test whether the diets A and B differ significantly as regards their effect on increase in weight.

Given the following :

Degrees of freedom:	*19*	*20*	*21*	*22*	*23*
Value of t at 5% level:	*2.09*	*2.09*	*2.08*	*2.07*	*2.07*

Solution:

Let us take the null hypothesis that A and B do not differ significantly weight regard to their effect on increase in weight. Applying t-test :

$$t = \frac{\overline{X}_1 - \overline{X}_2}{S}\sqrt{\frac{n_1 n_2}{n_1 + n_2}}$$

Calculating the required values :

Persons fed on diet A			Persons fed on diet B		
Increases in weight	Deviations from mean 12	$(X_1 - \overline{X}_1)^2$	Increase in weight	Deviations from actual mean 15	$(X_2 - \overline{X}_2)^2$
X_1	$(X_1 - \overline{X}_1)$		X_2	$(X_2 - \overline{X}_2)$	
10	–2	4	7	–8	64
6	–6	36	13	–2	4
16	+4	16	22	+7	49
17	+5	25	15	0	0
13	+1	1	12	–3	9
12	0	0	14	–1	1
8	–4	16	18	+3	9
14	+2	4	8	–7	49
15	+3	9	21	+6	36
9	–3	9	23	+8	64
			10	–5	25
			17	+2	4
$\Sigma X_1 = 120$	$\Sigma(X_1 - \overline{X}_1) = 0$	$\Sigma(X_1 - \overline{X}_1)^2 = 120$	$\Sigma X_2 = 180$	$\Sigma(X_2 - \overline{X}_2) = 0$	$\Sigma(X_2 - \overline{X}_2)^2 = 314$

Mean increase in weight of 10 persons fed on diet A

$$\overline{X}_1 = \frac{\Sigma X_1}{n_1} = \frac{120}{10} = 12 \text{ pounds}$$

Mean increase in weight of 12 persons fed on diet B

$$\overline{X}_2 = \frac{\Sigma X_2}{n_2} = \frac{180}{12} = 15 \text{ pounds}$$

$$S = \sqrt{\frac{\Sigma\left(X_1 - \overline{X}_1\right)^2 + \Sigma\left(X_2 - \overline{X}_2\right)^2}{n_1 + n_2 - 2}} = \sqrt{\frac{120 + 314}{10 + 12 - 2}} = \sqrt{\frac{434}{20}} = 4.66$$

$\overline{X}_1 = 12$, $\overline{X}_2 = 15$, $n_1 = 10$, $n_2 = 12$, $S = 4.66$. Substituting the values in the above formula :

$$t = \frac{12 - 15}{4.66} \times \sqrt{\frac{10 \times 12}{10 + 2}} = \frac{3}{4.66} \times 2.34 = 1.51$$

$v = n_1 + n_2 - 2 = 10 + 12 - 2 = 20.$

For v = 20, the table value of t at 5 per cent level is 2.09. The calculated value is less than the table value and hence the experiment provides no evidence against the hypothesis. We, therefore, conclude that diets A and B do not significantly as regards their effect on increase in weight is concerned.

MISCELLANEOUS EXAMPLES

Example 1:

How many pairs of observations must be included in a sample in order than an observed correlation coefficient of value 0.42 shall have a calculated value of t greater than 2.72?

Solution:

$$t = \frac{r}{\sqrt{1-r^2}} \times \sqrt{n-2}$$

We are given the value of t and r and we have to find out n.

$$\therefore \quad \frac{0.42}{\sqrt{1-(0.42)^2}} \times \sqrt{n-2} > 2.72$$

$$\text{or} \quad \frac{0.42}{0.908} \times \sqrt{n-2} < 2.72$$

$$\text{or} \quad 0.42 \times \sqrt{n-2} > 2.72 \times 0.908$$

$$\text{or} \quad \sqrt{n-2} > \frac{2.72 \times 0.908}{0.42} = 5.88$$

$$\text{or} \quad n - 2 > (5.88)^2 = 34.57$$

$$\text{or} \quad n - 24.57 + 2 = 36.57 \text{ or } 37$$

Hence we should include 37 observations.

Example 2:

The mean life of a sample of 10 electric light bulbs was found to be 1,456 hours with standard deviation of 423 hours. A second sample of 17 bulbs chosen from a different batch showed a mean life of 1,280 hours with standard deviation of 398 hours. Is there a significant difference between the means of the two batches?

Solution:

Let us take the hypothesis that the means of two batches do not differ significantly. Applying t-test :

$$t = \frac{\overline{X}_1 - \overline{X}_2}{S}\sqrt{\frac{n_1 n_2}{n_1 + n_2}}$$

$\overline{X}_1 = 1456,\ n_1 = 10,\ S_1 = 423,$

$\overline{X}_2 = 1280,\ n_2 = 17,\ S_2 = 398$

$$S = \sqrt{\frac{(n_1 - 1)\ S_1^2 + (n_2 - 1)\ S_2^2}{n_1 + n_2 - 2}} = \sqrt{\frac{9(423)^2 + 16(398)^2}{10 + 17 - 2}}$$

$$= \sqrt{\frac{1610361 + 2534464}{25}} = \sqrt{165793} = 407.18$$

$$t = \frac{1456 - 1280}{407.18}\sqrt{\frac{10 \times 17}{10 + 17}} = \frac{176}{407.18} \times 2.51 = 1.085$$

$v = 10 + 17 - 2 = 25;$

For $v = 25,\ t_{0.05} = 2.06$

The calculated value of t is less than the table value. Hence the hypothesis holds true. We, therefore, conclude that the means of two batches do not differ significantly.

Example 3:

The following table gives the ages in years of 10 husbands and their wives at marriage. Compute the correlation coefficient and test for its significance.

Solution:

We set up the hypothesis that there is no correlation in the population, Applying t-test

$$t = \frac{r}{\sqrt{1 - r^2}} \times \sqrt{n - 2} = \frac{0.939}{\sqrt{1 - (0.939)^2}} \times \sqrt{10 - 2}$$

$$= \frac{0.939}{\sqrt{1 - 0.882}} \times 2.8284 = \frac{0.939}{0.344} \times 2.8284 = 7.72$$

$v = n - 2 = (10 - 2) = 8$

For $\quad v = 8,\ t_{0.05} = 2.31$

The calculated value being much higher than the table value, the correlation is significant.

Example 4:

In a test given to two groups of students, the marks obtained are as follows :

First Group : 18 20 36 50 49 36 34 49 41

Second Group : 29 28 26 35 30 44 46

Examine the significance of difference between the arithmetic mean of the marks secured by the students of the above two groups.

(The value of t at 5% level of significance for v = 14 is 2.14)

Solution:

Let us take the hypothesis that there is no significant difference in the arithmetic mean of the marks secured by the students of the two groups. Applying t-test :

$$t = \frac{\overline{X}_1 - \overline{X}_2}{S}\sqrt{\frac{n_1 n_2}{n_1 + n_2}}$$

Group I X_1	$(X_1 - \overline{X}_1)$ $\overline{X}_1 = 37$	$(X_1 - \overline{X}_1)^2$	Group II X_2	$(X_2 - \overline{X}_2)$ $\overline{X}_2 = 34$	$(X_2 - \overline{X}_2)^2$
18	– 19	361	29	–5	25
20	– 17	289	28	– 6	36
36	– 1	1	26	– 8	64
50	+ 13	169	35	+ 1	1
49	+ 12	144	30	– 4	16
36	– 1	1	44	+ 10	100
34	– 3	9	46	+ 12	144
49	+ 12	144			
41	+ 4	16			
$\Sigma X_1 = 333$	$\Sigma(X_1 - \overline{X}_1) = 0$	$\Sigma(X_1 - \overline{X}_1)^2 = 1{,}134$	$\Sigma X_2 = 238$	$\Sigma(X_2 - \overline{X}_2) = 0$	$\Sigma(X_2 - \overline{X}_2)^2 = 386$

$$\overline{X}_1 = \frac{\Sigma X_1}{n_1} = \frac{333}{9} = 37; \ \overline{X}_2 = \frac{\Sigma X_2}{n_2} = \frac{238}{7} = 34$$

$$S = \sqrt{\frac{\Sigma\left(X_1 - \overline{X}_1\right)^2 + \Sigma\left(X_2 - \overline{X}_2\right)^2}{n_1 + n_2 - 2}} = \sqrt{\frac{1134 + 386}{9 + 7 - 2}} = 10.42$$

$$t = \frac{37 - 34}{10.42}\sqrt{\frac{9 \times 7}{9 + 7}} = \frac{3}{10.42} \times 1.984 = 0.571.$$

$v = n_1 + n_2 - 2 = 9 + 7 - 2 = 14;$

For $v = 14$, $t_{0.05} = 2.14$

The calculated value of t is less than the table value and hence the hypothesis holds true. We, therefore, conclude that the mean marks of the students of the two groups do not differ significantly.

Example 5:

Test the significance of the correlation r = 0.5 from a sample of size 18 against hypothesis correlation r = 0.7.

Solution:

We have to test the hypothesis that correlation in the population is 0.7.

Applying Z-transformation

$$Z = \frac{1}{2}\log_e \frac{1+r}{1-r} = 1.1513 \log_{10} \frac{1+0.5}{1-0.5}$$

$= 1.1513 \log 3 = 1.1513 \times 0.4771 = 0.549$

$$\xi = \frac{1}{2}\log_e \frac{1+\rho}{1-\rho} = \frac{1}{2}\log_{10} \frac{1+0.7}{1-0.7}$$

$= 1.1513 \log 5.67 = 1.1513 \times 0.7536 = 0.868.$

$$S.E._z = \frac{1}{\sqrt{n-3}} = \frac{1}{\sqrt{15}} = \frac{1}{3.873} = 0.258$$

$$\frac{\text{Difference}}{\text{S.E.}} = \frac{0.319}{0.258} = 1.236$$

Since the difference is less than 1.96 S.E. (5% level of significance) it could have arisen due to fluctuations of sampling. Hence ρ may very well be 0.7.

Example 6:

The heights of six randomly chosen soldiers are inches : 76, 70, 68, 69, 89, and 68. Those of 6 randomly chosen sailors are 68, 64, 69, 72, 74. Discuss the light the these data throw on the suggestions that soldiers are, on the average, tailor man sailors. Use t-test :

Solution:

Let us take the hypothesis that there is no difference is height of soldiers and sailors. Applying t-test :

$$t = \frac{\overline{X}_1 - \overline{X}_2}{S}\sqrt{\frac{n_1 n_2}{n_1 + n_2}}$$

Height X_1	$(X_1-\overline{X}_1)$	$(X_1-\overline{X}_1)^2$	Height X_2	$(X_2-\overline{X}_2)$	$(X_2-\overline{X}_2)^2$
		= 46			
76	+6	36	68	+1	1
70	0	0	64	–3	9
68	–2	4	65	–2	4
69	–1	1	69	+2	4
69	–1	1	72	+5	25
68	–2	4	64	–34	9
		$\Sigma(X_1-\overline{X}_1)^2$	$\Sigma X_2 = 402$		$\Sigma(X_2-\overline{X}_2)^2 = 52$

$$\overline{X}_1 = \frac{420}{6} = 70;\ \overline{X}_2 = \frac{402}{6} = 67$$

$$S = \sqrt{\frac{\Sigma\left(X_1 - \overline{X}_1\right)^2 + \Sigma\left(X_2 - \overline{X}_2\right)^2}{n_1 + n_2 - 2}}$$

$$= \sqrt{\frac{46+52}{6+6-2}} = \sqrt{\frac{98}{10}} = 3.13$$

$$= \frac{70-67}{3.13}\sqrt{\frac{6\times 6}{6+6}} = \frac{3}{3.13}\times 1.732 = 1.66$$

$$v = n_1 + n_2 - 2 = 6 + 6 - 2 = 10$$

For $v = 10,\ t_{0.05} = 2.23.$

The calculated value of t is less than the table value. There is no reason to doubt the hypothesis. Hence the soldiers are not, on an average, taller than sailors.

Example 7:

Out of 20,000 customer's ledger accounts, a sample of 600 was taken to test the accuracy of posting and balancing and 45 mistakes were found. Assign limits which the number of mistakes can be expected at 95% level of confidence.

Solution:

p, *i.e.*, Proportion of mistakes = 45/600 = .075

$\therefore \quad q = 1 - .075 = .925$

$$S.E._p = \sqrt{\frac{pq}{n}} = \sqrt{\frac{.075 \times .925}{600}} = .011$$

95% Confidence Limits

p ± 1.96 S.E.

.075 ± 1.96 (.011)

.075 ± .022 = .053 to .097.

Number of mistakes = 2000 × 0.053 to 2000 × 0.097

= 1060 to 1940

Hence at 5% level of significance it is expected that the number of mistakes would vary between 5.3 to 9.7 per cent.

Example 8:

A factory is producing 50,000 pairs of shoes daily. From a sample of 500 pairs, 2% were found to be of sub-standard quality. Estimate the number of pairs that can be reasonably expected to be spoiled in the daily production and assign limits at 95% level of confidence.

Solution:

p = 0.02, q = 0.98

$$S.E._p = \sqrt{\frac{pq}{n}} = \sqrt{\frac{0.2 \times .98}{500}} = 0.0063$$

95% confidence limit for percentage of defective pairs of shoes is given by

p ± 1.96 S.E.

0.2 ± 1.96 (.0063)

.02 ± .0123 = .0077 to .0323

The number of pairs that can be reasonably expected to be spoiled in daily production is given by 50,000 × .0077, *i.e.*, 385 and 50,000 × 1615.

Example 9:

10 workers are selected at random from a large number of workers in a factory. The number of items produced by them on a certain day are found to be:

51 52 53 55 56 57 58 59 59 60

In the light of these data, would it be appropriate to suggest that the mean of the number of items produced in the population is 58? (5% value of t for 9 d.f. is 2.262).

Solution:

Let us first calculate the sample mean and standard deviation

X	(X – 56)	$(X - 56)^2$
51	–5	25
52	–4	16
53	–3	9
55	–1	1
56	0	0
57	+1	1
58	+2	4
59	+3	9
59	+3	9
60	+4	16
ΣX = 560		$\Sigma(X - 56)^2 = 90$

$$\overline{X} = \frac{\Sigma X}{n} = \frac{560}{10} = 56$$

$$S = \sqrt{\frac{\Sigma\left(X - \overline{X}\right)^2}{n-1}} = \sqrt{\frac{90}{6}} = 3.162$$

Let us take the hypothesis that there is no significant difference in the sample mean and hypothetical population mean. Applying t-test.

$$t = \frac{\overline{X} - \mu}{S}\sqrt{n}$$

$\overline{X} = 56$, m = 58, S = 3.162, n = 10 .

$$t = \frac{|56 - 58|}{3.162} \times \sqrt{10} = \frac{2}{3.162} \times 3.162 = 2$$

v = 10 – 1 = 9.

For v 9 $t_{0.05} = 2.262$

The calculated value of t is less than the table value. The hypothesis is accepted. Hence in the light of the data it is appropriate to suggest that the mean of the number of items produced in the population is 58.

Example 10:

In a village 'A' out of a random sample of 1,000 persons 100 were found to be vegetarians while in another 'B' out 1,500 persons 180 were found to be vegetarians. Do you find a significant difference in the food habits of the people of the two villages?

Solution:

Let us take the hypothesis that there is no significant difference in the food habits of the people of two villages. Applying the test of the difference of two proportions; $S.E._{(p_1-p_2)} = \sqrt{pq\left(\frac{1}{n_1}+\frac{1}{n_2}\right)}$

p_1, *i.e.*, percentage of vegetarians in village 'A' = $\frac{100}{1000} \times 100 = 10$

p_2, *i.e.*, percentage of vegetarians in village 'B' = $\frac{180}{1500} \times 100 = 12$

$$p = \frac{x_1 + x_2}{n_1 + n_2} = \frac{100 + 180}{1000 + 1500} = \frac{280}{2500} = 0.112 \text{ or } 11.2 \text{ per cent}$$

$q = 100 - 11.2 = 88.8\%$

$$S.E._{(p_1-p_2)} = \sqrt{11.2 \times 88.8\left(\frac{1}{1000}+\frac{1}{1000}\right)} = \sqrt{\frac{994.56 \times 5}{3000}} = 1.288$$

$$\frac{\text{Difference}}{S.E.} = \frac{12-10}{1.288} = 1.55$$

Since the calculated value (1.56) is less than 1.96 at 5% level of significance there is no reason to doubt the hypothesis. Hence is no significant difference in the food habits of the people of two villages 'A' and 'B'.

Example 11:

An educator claims that the average I.Q. of American college students is at most 110, and that in a study made to test this claim 150 American college students, selected at random, had an average I.Q. of 111.2 with a standard deviation of 7.2. Use a level of significance of 0.01 to test the claim of the educator.

Solution:

Let us take the hypothesis that there is no significant difference in the claim of the educator and the sample result.

$$\frac{\text{Diff.}}{\text{S.E.}\bar{x}} = \frac{111.2 - 110}{0.588} = 2.04$$

Since the difference is less than 2.58 S.E. (1% level of significance), the hypothesis is accepted. Hence the claim of the educator is valid.

Example 12:

The following data give samples sizes and correlation coefficients. Test the significance of the difference between two values using Fisher's Z-transformation.

Sample Size	*Value of r*
5	0.870
12	0.560

Solution:

Let us take the hypothesis that the samples are drawn from the same population. Applying Z-test

$$Z = \frac{Z_1 - Z_2}{\sqrt{\dfrac{1}{n_1 - 3} + \dfrac{1}{n_2 - 3}}}$$

$$Z_1 = \log Z_1 = \log_e \frac{1 + r_1}{1 - r_1} \text{ or } 1.1513 \ \log_{10} \frac{1 + r_1}{1 - r_1}$$

Here $Z_1 = 1.1513 \log \dfrac{1 + 0.87}{1 - 0.87}$

$$= 1.1513 \log \frac{1.87}{0.13}$$

$$= 1.1513 \log 14.385$$

$$= 1.1513 \times 1.1579 = 1.333$$

$$Z_2 = \frac{1}{2} \log_e \frac{1 + r_2}{1 - r_2} \text{ or } 1.1513 \ \log_{10} \frac{1 + r_2}{1 - r_2}$$

$$= 1.1513 \log_{10} \frac{1 + 0.56}{1 - 0.56}$$

$$= 1.1513 \log \frac{1.56}{0.44}$$

$$= 1.1513 \log 3.545$$

$$= 1.1513 \times 0.5496 = 0.633$$

$$Z_1 - Z_2 = 1.333 - 0.633 = 0.7$$

$$S.E._{(Z_1 - Z_2)} = \sqrt{\frac{1}{n_1 - 3} + \frac{1}{n_2 - 3}}$$

$$= \sqrt{\frac{1}{5-3} + \frac{1}{12-3}} = 0.782$$

$$\frac{Z_1 - Z_2}{S.E.} = \frac{0.7}{0.782} = 0.895.$$

Since the difference is less than 2.58 S.E. (1% level) the experiment provides no evidence against the hypothesis that the samples are drawn from the sample population.

Example 13:

A population consists of four numbers 3, 7, 11, 15. Consider all possible samples of size two which can be drawn with replacement from their population. Find

(i) *the population mean,*

(ii) *the population standard deviation,*

(iii) *the mean of the sampling distribution of means, and*

(iv) *the standard deviation of the sampling distribution of means.*

Verify (iii) and (iv) directly from (i) and (ii) by use of suitable formulae (which you are to mention). Solve this problem if sampling is without replacement.

Solution:

With 4 numbers 16 samples (*i.e.* 4 × 4) of size two can be drawn with replacement. The samples with their respective means are shown below:

Sample	Mean	Sample	Mean	Sample	Mean	Sample	Mean
3, 3	3	3, 7	5	3, 11	7	3, 15	9
7, 3	5	7, 7	7	7, 11	9	7, 15	11
11, 3	7	11, 7	9	11, 11	11	11, 15	13
15, 3	9	15, 7	11	15, 11	13	15, 15	15
	24		32		40		48

Mean of the sampling distribution of means :

$$= \frac{24+32+40+48}{16} = 9$$

The mean of the population $= \frac{3+7+11+15}{4} = 9$

Hence the mean of the sampling distribution of means is equal to the population mean.

Standard Deviation of Sampling Distribution

Mean	Deviations (d)	d^2	Mean	Deviations (d)	d^2
3	(3 – 9) = – 6	36	7	(7 – 9) = – 2	4
5	(5 – 9) = – 4	16	9	(9 – 9) = 0	0
7	(7 – 9) = – 2	4	11	(11 – 9) = 2	4
9	(9 – 9) = 0	0	13	(13 – 9) = 4	16
5	(5 – 9) = – 4	16	9	(9 – 9) = 0	0
7	(7 – 9) = – 2	4	11	(11 – 9) = 2	4
9	(9 – 9) = 0	0	13	(13 – 9) = 4	16
11	(11 – 9) = 2	4	15	(15 – 9) = 6	36
		$\Sigma d^2 = 80$			$\Sigma d^2 = 80$

S.D. of sampling distribution $= \sqrt{\frac{160}{16}} = 3.162$

Standard Deviation of Population

X	(X – 9) x	x^2
3	– 6	36
7	–2	4
11	+ 2	4
15	+ 6	36
	$\Sigma x = 5$	$\Sigma x^2 = 80$

$$\sigma = \sqrt{\frac{\Sigma x^2}{n}} = \sqrt{\frac{80}{4}} = 4.472$$

The standard deviation of sampling distribution of means is equal to standard error of mean, *i.e.*, $\frac{\sigma}{\sqrt{n}} = \frac{4.47}{\sqrt{2}} = 3.16$ (n = size of sample).

If the sampling is without replacement the following six samples can be drawn :

(3, 7) (3, 11), (3, 15), (7, 11), (7, 15), (11, 15).

Their respective means are 5, 7, 9, 9, 11, 13

Mean of sampling distribution of mean

$$= \frac{5 + 7\ 9 + 9 + 11 + 13}{6} = \frac{54}{6} = 9$$

It is the same as population mean.

Variance of sampling distribution of means

$$= \frac{(5-9)^2 + (7-9)^2 + (9-9)^2 + (9-9)^2 + (11-9)^2 + (13-9)^2}{6}$$

$$= \frac{16+4+4+16}{6} = \frac{40}{6} = 6.67$$

$$\sigma = \sqrt{6.67} = 2.58$$

It is equal to

$$\sqrt{\left(\frac{n-n_1}{n-1}\right)\frac{\sigma^2}{\sqrt{n_1}}} = \sqrt{\left(\frac{4-2}{4-1}\right)\frac{20}{3}} = \sqrt{\frac{20}{3}} = \sqrt{6.67} = 2.58$$

Example 14:

A sample of 100 households in a village was taken and the average income was found to be Rs. 628 per month with a standard deviation of Rs. 60 per month. Find the standard error of mean and determine 99% confidence limits within which the incomes of all the people in this village are expected to lie.

Solution:

$$\text{S.E.}\left(\overline{X}\right) = \frac{\sigma}{\sqrt{n}} = \frac{50}{\sqrt{100}} = \frac{50}{10} = 5$$

99% Confidence Limits

$\overline{X} \pm 2.58$ S.E. $= 628 \pm 2.58(5)$

$= 628 \pm 12.9$ or 615.1 to 640.9

Hence the limits which the incomes of all the people in this village are expected to be are Rs. 615.1 to Rs. 649.9.

Example 15:

A sample of size 25 yielded mean equal 1 to 33 and an estimated variance equal to 100. At the 1% level would we have reasons to doubt the claim that the population mean is not greater than 27?

Solution:

Let us take the hypothesis that there is no significant difference between the sample mean and the hypothetical population mean. Applying t-test.

$$t = \frac{\overline{X} - \mu}{S}\sqrt{n}$$

$\overline{X} = 33, \mu = 27, S = \sqrt{100} = 10, n = 25.$

$$t = \frac{33-27}{10} \times \sqrt{25} = 3$$

$v = n - 1 = 25 - 1 = 24$

For $v = 24$, $t_{0.05} = 2.064$.

The calculated value of t is more than the table value. The hypothesis is rejected. Hence we have reasons to doubt the claim that the population mean is not greater than 27.

Example 16:

The mean population of a random sample of 400 villages in Jaipur district was found to be 400 with a standard deviation of 12. The mean population of a random sample of 400 villages in Meerut district was found to be 395 with a standard deviation of 15, is the difference between the two means statistically significant.

$$S.E.\ \left(\overline{X}_1 - \overline{X}_2\right) = \sqrt{\frac{\sigma_1^2}{n_1} + \frac{\sigma_2^2}{n_2}}$$

$\sigma_1 = 12,\ n_1 = 400,\ \sigma_2 = 15,\ n_2 = 400,$

$$S.E. = \sqrt{\frac{(12)^2}{400} + \frac{(15)^2}{400}} = \sqrt{\frac{144}{400} + \frac{225}{400}}\ 0.96$$

$\overline{X}_1 - \overline{X}_2 = 400 - 395 = 5$

$$\frac{\text{Difference}}{\text{S.E}} \; \frac{5}{9.6} = 5.21$$

Since the difference is more than 2.58 (1% level of significance), the hypothesis is rejected. Hence the difference between the mean population of the two villages is statistically significant.

Example 17:

In measuring reaction time, a psychologist estimates that the standard deviation is 0.05 seconds. How large a sample of measurements must be taken in order to be 95% confident that the error of his estimate will not exceed 0.001 seconds?

Solution:

Since the error of estimate is not to, exceed 0.01 second

$$\frac{1.96\,\sigma}{\sqrt{n}} > 0.01$$

$$1.96 \times 0.05 > 0.01\sqrt{n}$$

or $$\sqrt{n} < \frac{1.96 \times 0.05}{0.01} = 9.8$$

$$n \leq (9.8)^2 = 96.04.$$

Hence the size of the sample should be if.

Example 18:

An examination was given to two classes consisting of 40 and 50 students respectively. In the first class the mean mark was 74 with a standard deviation of 8, while in the second class the mean mark was 78 with a standard deviation of 7. Is there a significant difference between the performance of the two classes at a level of significance of 0.05?

Solution:

Let us take the hypothesis that there is no significant difference in the mean marks of the two classes.

$$\text{S.E.}\left(\overline{X}_1 - \overline{X}_2\right) = \sqrt{\frac{\sigma_1^2}{n_1} + \frac{\sigma_2^2}{n_2}}$$

$\sigma_1 = 8$, $n_1 = 40$, $\sigma_2 = 7$, $n_2 = 50$

$$\text{S.E.} = \sqrt{\frac{(8)^2}{40} + \frac{(7)^2}{50}} = \sqrt{1.6 + 0.98}\ 1.606$$

$$\frac{\text{Difference}}{\text{S.E.}} = \frac{78 - 74}{1.606} = 2.49$$

Since the difference is more than 1.96 S.E. (5% level) of significance, the hypothesis is rejected. Hence there is a significant difference in the performance of the two classes at 5% level.

Example 19:

An I.O. test was administered to 5 persons before and after they were trained. The results are given below :

Candidates	***I***	***II***	***III***	***IV***	***V***
I.Q. before training	*110*	*120*	*123*	*132*	*125*
I.Q. after training	*120*	*118*	*125*	*136*	*121*

Test whether there is any change I.Q. after the training programme.

[For v = 4, $t_{0.01} = 4.6$*]*

Solution:

Let us take the hypothesis that there is no change in I.Q. after the training programme. Applying t-test :

$$t = \frac{\bar{d}\sqrt{n}}{S}$$

I.Q. Before	**I.Q. After**	**d**	**d²**
110	120	+ 10	100
120	118	– 2	4
123	125	+ 2	4
132	136	+ 4	16
125	121	– 4	16
		$\Sigma d = 10$	$\Sigma d^2 = 140$

$$\bar{d} = \frac{\Sigma d}{n} = \frac{10}{5} = 2$$

$$S = \sqrt{\frac{\Sigma d^2 - n(\bar{d})^2}{n-1}} = \sqrt{\frac{140 - 5(2)^2}{5-1}} = 5.477$$

$$t = \frac{2\sqrt{5}}{5.477} = 0.817$$

For v = 4, $t_{0.01} = 4.6$

The calculated value of t is less than the table value. The hypothesis holds true. Hence there is no change in I.Q. after the training programme.

Example 20:

Two randomly selected groups of 50 employees each of a very large firm are taught an assembly operation by two different methods and then tested for performance if the first group average 140 points with a standard deviation of 10 points while the second group averaged 135 points with a standard deviation of 8 points, test at 0.05 level whether the difference between their mean scores significant.

Solution:

Let us take the hypothesis that the difference in the mean scores of the two groups of employees is not significant. Applying t-test of difference of means.

$$\text{S.E. } (\overline{X}_1 - \overline{X}_2) = \sqrt{\frac{\sigma_1^2}{n_2} + \frac{\sigma_2^2}{n_2}}$$

$\sigma_1 = 10$, $\sigma_2 = 8$, $n_1 = 50$, $n_2 = 50$, $\overline{X}_1 = 140$ $\overline{X}_2 = 135$

$$= \sqrt{\frac{(10)^2}{50} + \frac{(8)^2}{50}} = \sqrt{2 + 128} = 1.811$$

$$\frac{\text{Diff.}}{\text{S.E.}} = \frac{140 - 135}{1.811} = 2.76$$

Since the difference is more than 1.96 S.E. at (5% level) of significance, the hypothesis is rejected. Hence the difference in the mean scores of the two groups of employees is significant.

Example 21:

You are working as a purchase manager for a company. The following information has been supplied to you by two manufacture of electric bulbs:

	Company A	*Company B*
Mean life in hours	*1300*	*1248*
Standard deviations (in hours)	*82*	*93*
Sample size	*100*	*100*

Which brand of bulbs are you going to purchase if you desire to take a risk of 5%?

Solution:

Let us take the hypothesis that there is no significant difference in the mean life of the two makes of bulbs.

$$\text{S.E.}\left(\overline{X}_1 - \overline{X}_2\right) = \sqrt{\frac{\sigma_1^2}{n_1} + \frac{\sigma_2^2}{n_2}}$$

$\sigma_1 = 82, \sigma_2 = 93, n_1 = 100, n_2 = 100$

$$\text{S.E.}_{(\overline{X}_1 - \overline{X}_2)} = \sqrt{\frac{(82)^2}{100} + \frac{(93)^2}{100}} = \sqrt{\frac{6724}{100} + \frac{8649}{100}}$$

$$\frac{\text{Diff.}}{\text{S.E.}} = \frac{1300 - 1248}{12.399} = 4.19$$

Since the difference is more than 1.96 S.E., the hypothesis is rejected. Hence there is a significant difference in the mean life of the two makes of bulbs. We should prefer to buy the bulbs of Co. A since their average life is more.

Example 22:

(a) 400 labourers were selected at random from a certain district. Their mean income was 140.5 rupees per month, with standard deviation 25.2 rupees. Set up 95% confidence limits within which the income of the labour community of district is expected to lie.

(b) Set up 95% confidence limits for the standard deviation of the entire labour community of the district.

Solution:

(a) We are given = 140.5, σ = 25.2 and n = 400

$$E\overline{x} = \frac{\sigma}{\sqrt{n}} = \frac{25.2}{\sqrt{400}} = 1.26$$

95% confidence limits for mean are given by :

$\overline{X} \pm 1.96$ S.E.

$= 140.5 \pm 1.96\ (1.26)$

$= 140.5 \pm 2.47$ or 138.03 to 142.97

(b) $S.E._{\sigma} = \frac{\sigma}{\sqrt{2n}} = \frac{25.2}{\sqrt{2 \times 400}} = 0.89$

95% confidence limit for standard deviation are given by :

$\sigma + 1.96\ S.E._{\sigma}$

or $25.2 \pm 1.96\ (.89)$

25.2 ± 26.944

or 23.456 to 26.944.

Example 23:

A sample of 400 male students is found to have a mean height of 171.38 cm. Can if be reasonably regarded as a sample from a large population with mean height 171.17 cm. and standard deviation 3.30 cm.?

Solution:

Let us take the hypothesis that there is no significant in the sample mean and the population mean. Calculating standard error of mean :

$$S.E.\bar{x} = \frac{\sigma}{\sqrt{n}} = \frac{3.3}{400} = \frac{3.3}{20} = 0.165$$

$$\frac{\text{Diff.}}{S.E.} = \frac{171.38 - 171.17}{0.165} = 1.273$$

Since the difference is less than 1.96 S.E. (5% level) the hypothesis is accepted. Hence there is no significant difference in the sample mean and the population mean.

Example 24:

Sample of two different types of bulbs were tested for length of life, and the following data were obtained :

	Type I	***Type II***
Sample size	*8*	*7*
Sample mean	*1234 hrs*	*1136 hrs*
Sample S.D.	*36 hrs*	*40 hrs*

Is the difference in the means significant? (Given that the significant value of t at 5% level of significance for 13 d.f. is 2.16.).

Solution:

Let us the hypothesis that there is significant difference in mean life of the two makes of bulbs I and II. Applying t-test of the difference of means.

$$t = \frac{\overline{X}_1 - \overline{X}_2}{S}\sqrt{\frac{n_1 n_2}{n_1 + n_2}}$$

$$S = \sqrt{\frac{(n_1 - 1)\,S_1^2 + (n_2 - 1)\,S_2^2}{n_1 + n_2 - 2}}$$

$$= \sqrt{\frac{7(36)^2 + 6(40)^2}{8 + 7 - 2}} = 37.898$$

$\overline{X}_1 = 1234$, $\overline{X}_2 = 1136$, $S = 37.898$, $n_1 = 8$, $n_2 = 7$

$$t = \frac{1234 - 1136}{101.48}\sqrt{\frac{8 \times 7}{8 + 7}} = \frac{98 \times 1.932}{37.898} = 4.996$$

$v = 8 + 7 - 2 = 13$,

For $v = 13$, $t_{0.05} = 2.16$.

The calculated value of t is more than the table value. The hypothesis is rejected. Hence the difference in the mean is significant.

Example 25:

A college conducted both day and evening classes intended to be identical. A sample of 100 day students yields examination results as under:

$$\overline{X}_1 = 72.4 \quad \sigma_1 = 14.8$$

A sample of 200 evening students yields examination results as under:

$$\overline{X}_2 = 73.9, \quad \sigma_2 = 17.9.$$

Are the two statistically equal at 1% level?

Solution:

Let us take hypothesis that the two means are statistically equal. Calculating standard error of difference of means :

$$S.E.\left(\overline{X}_1 - \overline{X}_2\right) = \sqrt{\frac{\sigma_1^2}{n_1} + \frac{\sigma_2^2}{n_2}}$$

$$= \sqrt{\frac{(14.8)^2}{100} + \frac{(17.9)^2}{200}}$$

$$= \sqrt{2.1904 + 1.6021} = 1.947$$

$$\frac{\text{Diff.}}{\text{S.E.}} = \frac{|72.4 - 73.9|}{1.947} = 0.77$$

Since the difference is less than 2.58 S.E. (1% level of significance, the hypothesis holds true. Hence the two means may be regarded statistically equal).

$v = 12 + 7 - 2 = 17$

For $v = 17$, $t_{0.05} = 2.11$

The calculate value of t is more than the table value. The hypothesis rejected and we, therefore, conclude that the means of two samples differ significantly.

Example 26:

The following data were collected from two cities as regards the starting stipend paid to new management trainees.

Do the data give evidence that the stipend paid in city B is significantly more than that in city A?

Test at a significance level of 1%.

City	*Monthly Stipend (Mean)*	*Sample Standard (Deviation)*	*Sample Size*
A	*Rs. 1,400*	*Rs. 80*	*200*
B	*Rs. 1,600*	*Rs. 120*	*175*

Solution:

Let us take the hypothesis that there is no significant difference in the stipend paid in the two cities 'A' and 'B'. Applying large sample test of difference of means :

$$\text{S.E.}\left(\overline{X}_1 - \overline{X}_2\right) = \sqrt{\frac{\sigma_1^2}{n_1} + \frac{\sigma_2^2}{n_2}}$$

$\sigma_1 = 80$, $n_1 = 200$, $\sigma_2 = 120$, $n_2 = 175$.

$$\text{S.E.} = \sqrt{\frac{(80)^2}{200} + \frac{(120)^2}{175}} = \sqrt{32 + 82.286} = 10.69$$

$$\frac{\text{Diff.}}{\text{S.E.}} = \frac{|1400 - 1600|}{10.69} = 18.71$$

Since the difference is more than 2.58 S.E. (1% level of significance) the hypothesis is rejected. Hence there is significance in the stipend paid in the two cities.

Example 27:

Two independent samples of 8 and 7 items gave the following values:

Sample A :	*9*	*11*	*13*	*11*	*15*	*9*	*12*	*14*
Sample B :	*10*	*12*	*10*	*14*	*9*	*8*	*10*	

Examine whether the difference between the means of the two samples is significant at 5% level?

Solution:

Let us take the hypothesis that there is no significant difference in the means of the two samples. Applying t-test of difference of means,

$$t = \frac{\overline{X}_1 - \overline{X}_2}{S}\sqrt{\frac{n_1 n_2}{n_1 + n_2}}$$

X_1	$(X_1 - 12)$	$(X_1 - 12)^2$	X_2	$(X_2 - 10)$	$(X_2 - 10)^2$
9	− 3	9	10	0	0
11	− 1	1	12	+ 2	4
13	+ 1	1	10	0	0
11	− 1	1	14	+ 4	16
15	+ 3	9	9	− 1	1
9	− 3	9	8	− 2	1
12	0	0	10	0	0
14	+ 2	4	—	—	—
$\Sigma X_1=94$	$\Sigma(X_1 - A_1) = -2$	$\Sigma(X_1 - A_1)^2 = 34$	$\Sigma X_2 = 73$	$\Sigma(X_2 - 10) = 3$	$\Sigma(X_2 - 10)^2 = 25$

$$\overline{X}_1 = \frac{\Sigma X_1}{n_1} = \frac{94}{8} = 11.75;\ \overline{X}_2 = \frac{\Sigma X_1}{n_2} = \frac{73}{7} = 10.43$$

$$S = \sqrt{\frac{S(X_1 - A_1)^2 + S(X_2 - A_2)^2 - n_1\left(\overline{X}_1 - A_1\right)^2 - n_2\left(\overline{X}_2 - A_2\right)^2}{n_1 + n_2 - 2}}$$

$$= \sqrt{\frac{34+25-8(11.75-12)^2-7(10.43-10)^2}{8+7+2}}$$

$$= \sqrt{\frac{34+25-0.0625-0.1849}{13}} = \sqrt{4.519} = 2.126$$

$$t = \frac{11.75-10.43}{2.126}\sqrt{\frac{8 \times 7}{8+7}} = \frac{1 \cot 32}{2.126} \times 1.932 = 1.2$$

$v = n_1 + n_2 - 2 = 8 + 7 - 2 = 13$

For $v = 13$, $t_{0.05} - 2.16$

The calculated value of t is less than the table value. The hypothesis is accepted. Hence there is no significant difference in the means of the two samples.

Example 28:

For a random sample of size 10 from a normal population, the mean is 12.1 and the standard deviation is 3.2. It reasonable to suppose that the population mean is 14.5? Test at 5% significance level.

Solution:

Let us take the hypothesis that there is no significant difference between the sample mean and the population mean.

$H_0 = \overline{X} - \mu = 0$

$H_a = \overline{X} -- \mu \neq 0$

Applying t-test

$$t = \frac{\overline{X}-\mu}{S}\sqrt{n}$$

$\overline{X} = 12.1$, $\mu = 14.5$, $S = 3.2$, $n = 10$

$$t = \frac{|12.1-14.5|}{32} \times \sqrt{10} = 2.372$$

$v = n - 1 = 10 - 1 = 9.$

For $v = 9$, $t_{0.05} = 1.833$

The calculated value of t is greater than the table value. The hypothesis is rejected. Hence it is not reasonable to suppose that the population mean is 14.5.

Example 29:

Eleven sales executive trainees are assigned selling jobs right after their recruitment. After a fortnight they are withdrawn from their field duties and given a month's training for executive sales. Sales executed by them in thousands of rupees before and after the training, in the same period are listed below :

Sales ('000 Rs.) (Before Training) :

23 20 19 21 18 20 18 17 23 16 19

Sales ('000 Rs.) (After Training) :

24 19 21 18 20 22 20 20 23 20 27

Do these data indicate that the training has contributed to their performance?

Solution:

Let us take the hypothesis that the training has not contributed significantly to the performance of sales executives. Applying t-test :

Sales Before	Sales After	(2nd – 1st)	d^2
23	24	+ 1	1
20	19	– 1	1
19	21	+ 2	4
21	18	– 3	9
18	20	+ 2	4
20	22	+ 2	4
18	20	+ 2	4
17	20	+ 3	9
23	23	0	0
16	20	+ 4	16
19	27	+ 8	64
		$\Sigma d = 20$	$\Sigma d^2 = 116$

$$t = \frac{\bar{d}\sqrt{n}}{S}$$

$$\bar{d} = \frac{\Sigma d}{n} = \frac{20}{11} = 1.82$$

$$S = \sqrt{\frac{\Sigma d^2 - n(\bar{d})^2}{n-1}} = \sqrt{\frac{116 - 11\,(1.82)^2}{10}}$$

$$= \sqrt{\frac{116 - 36.44}{10}} = 2.82 \quad t = \frac{1.82\sqrt{11}}{2.82} = 2.14$$

$$v = n - 1 = 11 - 1 = 10$$

For $v = 10, t_{0.05} = 2.23$

The calculated value of t is less than the table value. Hence we accept the hypothesis. The training has not contributed significantly to the performance of sales executives.

Example 30:

The MFC Admission Test taken by two groups of boys and girls gave the following information :

	Mean score	*S.D.*	*Number*
Girls	*75*	*10*	*100*
Boys	*70*	*12*	*200*

On the basis of the above information would you conclude that mean scores of boys and girls are statistically significant?

Solution:

Let us take the hypothesis that there is no significant difference in the mean scores of boys and girls

$$S.E._{(\bar{X}_1 - \bar{X}_2)} = \sqrt{\frac{\sigma_1^2}{n_1} + \frac{\sigma_2^2}{n_2}} = \sqrt{\frac{(10)^2}{100} + \frac{(12)^2}{200}}$$

$$= \sqrt{\frac{100}{100} + \frac{144}{200}} = \sqrt{1 + 0.72} = 1.31. \frac{\text{Diff.}}{\text{S.E.}} = \frac{5}{1.31} = 3.82$$

Since the difference is more than 2.58 S.E. at 1% level of significance, the hypothesis is rejected. Hence the mean scores of boys and girls are statistically significant.

Example 31:

To test the desirability of a certain modification in typists, desks, 9 typists were given two test of as nearly as possible the same nature, one on the desk in use and the other on the new type. The following difference in the number of words type per minute recorded :

Typists :	*A*	*B*	*C*	*D*	*E*	*F*	*G*	*H*	*I*
Increase in number of words :	2	4	0	3	– 1	4	– 3	2	5

Do the data indicate the modification in desk promotes speed in typing?

Solution:

Let us take the hypothesis that the modification in desk does not promote speed in typing. Applying the t-test :

$$t = \frac{\bar{d}\sqrt{n}}{S}$$

Typist	**d**	**d²**
A	2	4
B	4	16
C	0	0
D	3	9
E	–1	1
F	4	16
G	–3	9
H	2	4
I	5	25
n = 9	**Σd = 16**	**Σd² = 84**

$$\bar{d} = \frac{\Sigma d}{n} = \frac{16}{9} = 1.778$$

$$S = \sqrt{\frac{\Sigma d^2 - n\left(\bar{d}\right)^2}{n-1}} = \sqrt{\frac{84-9\left(\frac{16}{9}\right)^2}{8}} = 2.635$$

$$t = \frac{1.778 \times 3}{2.635} = 2.635$$

v = 9 – 1 = 8. For v = 8, $t_{0.05}$ = 2.306

The calculated value of t is less than the table value and hence the hypothesis is accepted. The data does not indicate that the modification in desk promotes speed in typing.

Example 32:

Ten oil tins are taken at random from an automatic filing machine. The mean weight of the 10 tins is 15.8 kg. and standard deviation 0.5 kg. Does the sample mean differ significantly from the intended weight of 16 kg.? (given for $v = 9$, $t_{0.05} = 2.26$).

Solution:

Let us take the hypothesis that the sample mean does not differ significantly them the intended weight of 16 kg. Applying t-test :

$$t = \frac{\overline{X} - \mu}{S}\sqrt{n} \quad \overline{X} = 15.8,\ m = 16,\ S = 0.5,\ n = 16$$

$$\therefore \quad t = \frac{(15.8 - 16)}{0.5} = \sqrt{10} = \frac{2 \times 3.162}{0.5} = 1.27$$

$$v = n - 1 = 10 - 1 = 9$$

For $\quad v = 9,\ t_{0.05} = 2.26$

The calculated value of t is less than the table value. The hypothesis is accepted, *i.e.*, the sample mean does not differ significantly from the intended weight.

Example 33:

A random sample of 100 mill workers at Kanpur showed their mean wage to be Rs. 350 with a standard deviation of Rs. 28. Another random sample of 150 mill workers in Bombay showed the mean wage to be Rs. 390 with a standard deviation of Rs. 40. Do the mean wages of workers in Bombay and Kanpur differ significantly? Use .05 level of significance.

Solution:

Let us take the null hypothesis that there is no significant difference in the mean wages of workers in Bombay and Kanpur.

$$S.E._{(\overline{x}_1 - \overline{x}_2)} = \sqrt{\frac{\sigma_1^2}{n_1} + \frac{\sigma_2^2}{n_2}}$$

$$\sigma_1 = 28,\ n_1 = 100\ \sigma_2 = 40,\ n_2 = 150$$

$$S.E. = \sqrt{\frac{(28)^2}{100} + \frac{(40)^2}{150}} = \sqrt{84 + 10.67} = 4.3$$

$$\frac{\text{Diff.}}{\text{S.E.}} = \frac{390 - 350}{4 - 3} = 9.30$$

Since the difference is more than 1.96 S.E. at 5% level, the hypothesis is rejected. Hence the wages of workers in Bombay and Kanpur differ significantly.

Example 34:

A machine puts out 10 defective units in a sample of 200 units. After the machine is overhauled it puts out 4 defective units in a sample of 100 units. Has the machine been improved?

Solution:

Let us take the hypothesis that overhauling has not improved the performance of the machine. Applying the test of difference of two proportions, *i.e.*,

$$S.E._{(p_1-p_2)} = \sqrt{pq\left(\frac{1}{n_1}+\frac{1}{n_2}\right)}$$

$$p_1 = \frac{10}{200} = 0.05,\ p_2 = \frac{4}{100} = 0.04\,.$$

$$p = \frac{x_1+x_2}{n_1+n_2} = \frac{10+4}{200+100} = \frac{14}{300} = 0.47$$

$$q = 1 - .047 = 0.953$$

$$S.E. = \sqrt{(.047)\,(.953)\left(\frac{1}{200}+\frac{1}{100}\right)} = 0.026$$

$$\frac{\text{Diff.}}{\text{S.E.}} = \frac{.05-.04}{.026} - 0.385$$

Since the difference is less than 1.96 S.E. at 5% level of significance, the hypothesis is accepted.

Example 35:

A random sample of size 16 has 53 as mean. The sum of the squares of the deviations taken from mean is 135. Can this sample be regarded as taken from the population having 56 as mean? Obtain and 95% and 99% confidence limits of the mean of the population. (for $v = 15$, $t_{0.05} = 2.13$ for $v = 15$ $t_{0.01} = 2.95$)

Solution:

Let us take the hypothesis that there is no significant difference between the sample mean and hypothetical population mean. Applying t test:

$$t = \frac{\overline{X} - \mu}{S}\sqrt{n}$$

$\overline{X} = 53$, $\mu = 56$, $n = 16$, $S(X - \overline{X})^2 = 135$

$$S = \sqrt{\frac{\Sigma\left(X - \overline{X}\right)}{n-1}} = \sqrt{\frac{135}{15}} = 3$$

$$t = \frac{|53-56|}{3}\sqrt{16} = \frac{3 \times 4}{3} = 4$$

$v = 16 - 1 = 15.$

For $v = 16$, $t_{0.05} = 2.13$

The calculated value of t is more than the table value. The hypothesis is rejected. Hence the samples has not come from a population having 56 as mean.

95% confidence limits of population mean

$$\overline{X} \pm \frac{S}{\sqrt{n}} t_{0.05}$$

$$= 53 \pm \frac{3}{\sqrt{16}} \times 2.13$$

$$= 53 \pm 1.6 = 51.4 \text{ to } 54.6$$

99% confidence limits of the population mean

$$\overline{X} \pm \frac{S}{\sqrt{n}} t_{0.01}$$

$$= 53 \pm \frac{3}{\sqrt{16}} \times 2.95$$

$$= 53 \pm \frac{3}{4} \times 2.95$$

$$= 53 \pm 2.212 = 50.788 \text{ to } 55.212.$$

Example 36:

From a sample of 19 pairs of observations the correlation is 0.5 and the corresponding population value is 0.3. Is the difference significant?

Solution:

Using Z-transformation.

$$Z = \frac{1}{2} \log_e \frac{1+r}{1-r}$$

$$= 1.1513 \log_{10} \frac{1+r}{1-r} = 1.1513 \log_{10} \frac{1+0.5}{1-0.5}$$

$$= 1.1513 \log 3 = 1.1513 \times 0.4771 = 0.549$$

$$\xi = \log_{10} \frac{1+p}{1-\rho} \times 1.1513 = \log_{10} \frac{1+0.3}{1-0.3} \times 1.1513$$

$$= \log 1.857 \times 1.1513 = 0.2688 \times 1.1513 = 0.309$$

$$(Z - \xi) = 0.549 - 0.309 = 0.24$$

$$S.E._z = \frac{1}{\sqrt{n-3}} = \frac{1}{\sqrt{16}} = 0.25$$

$$\frac{\text{Difference}}{S.E.} = \frac{0.24}{0.25} = 0.96$$

Since the difference between Z and x is less than 2.58 S.E. (1% level) the difference could have arisen due to fluctuations of sampling. Hence the given correlation coefficient is not significantly different from 0.3.

To test the significance of the difference between two independent correlation coefficients. To test the significance of two correlation coefficients derived from two separate samples we have to compare the difference of the two corresponding values of Z with the standard error of that difference–remembering that the standard error of the difference of two statistical quantities is the square root of the sum of their variances. In other words, we have to apply the following formula :

$$Z = \frac{Z_1 - Z_2}{\sqrt{\frac{1}{n_1 - 3} + \frac{1}{n_2 - 3}}}$$

where $Z_1 = \frac{1}{2} \log_e \left(\frac{1+r_1}{1-r_1}\right)$ or $1.1513 \log_{10} \left(\frac{1+r_1}{1-r_1}\right)$

and $Z_2 = \frac{1}{2} \log_e \left(\frac{1+r_2}{1-r_2}\right)$ or $1.1513 \log_{10} \left(\frac{1+r_2}{1-r_3}\right)$

$$S.E._{(Z_1-Z_2)} = \sqrt{\frac{1}{n_1 - 3} + \frac{1}{n_2 - 3}}$$

If the absolute value of this statistic is greater than 1.96, the difference will be significant at 5% level. The following examples will illustrate the application of this test.

Example 37:

A drug is given to 10 patients, and the increments in their blood pressure were recorded to be 3, 6, –2, 4, –3, 4, 6, 0, 0, 2. Is it reasonable to believe that the drug has no effect on change of blood pressure? (5% value of t for 9 d.f. = 2.26).

Solution:

Let us take the hypothesis that the drug has no effect on change of blood pressure. Applying the difference test :

$$t = \frac{\bar{d}\sqrt{n}}{S}$$

d	**d²**
3	9
6	36
–2	4
4	16
–3	9
4	16
6	36
0	0
0	0
2	4
Σd = 0	**Σd² = 130**

$$\bar{d} = \frac{\Sigma d}{n} = \frac{20}{10} = 2$$

$$S = \sqrt{\frac{\Sigma d^2 - n\left(\bar{d}\right)^2}{n-1}} = \sqrt{\frac{130 - 10(2)^2}{10-1}} = 3.162$$

$$t = \frac{2\sqrt{10}}{3.162} = \frac{2 \times 3.162}{3.162} = 2$$

$v = n - 1 = 10 - 1 = 9$; For $v = 9$, $t_{0.05} = 2.26$.

The calculated value of t is less than the table value. The hypothesis is accepted. Hence it is reasonable to believe that the drug has no effect on change of blood pressure.

(4) *Testing the Significance of an Observed Correlation Coefficient:* Given a random sample from a bivariate normal population. If we are to test the hypothesis that the correlation coefficient of the population is zero, *i.e.*, the variables in the population are uncorrelated, we have to apply the following test :

$$t = \frac{r}{\sqrt{1-r^2}} \times \sqrt{n-2}$$

here t is based on (n – 2) degrees of freedom.

If the calculated value of t exceeds $t_{0.05}$ for (n – 2), d.f., we say that the value of r is significant at 5% level. If $t < t_{0.05}$ the data are consistent with the hypothesis of an uncorrelated population.

The following examples will illustrate the test:

Example 38:

Two types of drugs were used on 5 and 7 patients for reducing their drugs for six month was as follows :

Drug A :	*10*	*12*	*13*	*11*	*14*		
Drug B :	*8*	*9*	*12*	*14*	*15*	*10*	*9*

If the bias correlation due to small is ignored, pooled estimate of the standard deviation can be obtained by :

$$S = \sqrt{\frac{n_1 S_1^2 + n_2 S_2^2}{n_1 + n_2}}$$

Is there a significant difference in the efficacy of the two drugs? If not, which drug should you buy. (For v = 10, $t_{0.05}$ = 2.223)

Solution:

Let us take the hypothesis that there is no significant difference in the efficacy of the two drugs. Applying t-test :

$$t = \frac{\overline{X}_1 - \overline{X}_2}{S} \sqrt{\frac{n_1 n_2}{n_1 + n_2}}$$

X_1	$(X_1 - \overline{X}_1)$	$(X_1 - \overline{X}_1)^2$	X_2	$(X_2 - \overline{X}_2)$	$(X_2 - \overline{X}_2)^2$
10	–2	4	8	–3	9
12	0	0	9	–2	4
13	+1	1	12	+1	1
11	–1	1	14	+3	9
14	+2	4	15	+4	16
			10	–1	1
			9	–2	4
$\Sigma X_1 = 60$		$\Sigma(X_1 + \overline{X}_1)^2 = 10$	$\Sigma X_2 = 77$		$\Sigma(X_2 - \overline{X}_2)^2 = 44$

However, it is advisable to take account of bias.

$$\overline{X}_1 = \frac{\Sigma X_1}{n_1} = \frac{60}{5} = 12; \ \overline{X}_2 = \frac{\Sigma X_2}{n_2} = \frac{77}{7} = 11$$

$$S = \sqrt{\frac{\Sigma(X_1 - \overline{X}_1)^2 + \Sigma(X_2 - \overline{X}_2)^2}{n_1 + n_2 - 2}} = \sqrt{\frac{10+44}{5+7-2}} = \sqrt{\frac{54}{10}} = 2.324$$

$$t = \frac{\overline{X}_1 - \overline{X}_2}{S}\sqrt{\frac{n_1 n_2}{n_1 + n_2}}$$

$$\frac{12-11}{2.324}\sqrt{\frac{5\times7}{5+7}} = \frac{1.708}{2.324} = 0.735$$

$v = n_1 + n_2 - 2 = 5 + 7 - 2 = 10.$

For $v = 10, t_{0.05} = 2.228.$

The calculated value of t is less than the table value, the hypothesis is accepted. Hence there is no significance in the efficacy of two drugs. Since drug B is indigenous and there is no difference in the efficacy of imported and indigenous drug, we should buy indigenous drug, *i.e.*, B.

$$S = \sqrt{\frac{n_1 S_1^2 + n_2 S_2^2}{n_1 + n_2}}.$$

Example 39:

A company is interested in finding out if there is any difference in the average salary received by managers of two divisions. Accordingly samples of 12 managers in the first division and 10 managers in the second division were selected at random. The results are given below :

	I division	*II division*
Sample size	*12*	*10*
Average monthly salary	*12500*	*11200*
Standard deviation	*320*	*480*

Apply t-test to find out whether there is a significant difference in the average salary.

Solution:

Let us take the null hypothesis that there is no significant difference in the average salary of the two divisions. Applying t-test of difference of means.

$$t = \frac{\overline{X}_1 - \overline{X}_2}{S} \sqrt{\frac{n_1 n_2}{n_1 + n_2}}$$

$$S = \sqrt{\frac{(n_1 - 1)\, S_1^2 + (n_2 - 1)\, S_2^2}{n_1 + n_2 - 2}}$$

$$= \sqrt{\frac{(12-1)\,(320)^2 + (10-1)\,(480)^2}{12+10-2}}$$

$$= \sqrt{\frac{1126400 + 2073600}{20}} = 400$$

$\overline{X}_1 = 12500,\ \overline{X}_2 = 11200$

$S = 400,\ n_1 = 12,\ n_2 = 10$

$$\text{Hence } t = \frac{12500 - 11200}{400} \sqrt{\frac{12 \times 10}{12+10}}$$

$$= \frac{1300}{400} \times 2.335 = 7.59$$

$v = n_1 + n_2 - 2 = 12 + 10 - 2 = 10$

For $v = 10$, $t_{0.05} = 2.228$

The calculated value of t is more than the table value. The hypothesis is rejected. Hence there is a significant difference in the average salary of the two divisions.

In testing statistical significance the following points must be noted:

(1) *Non-publication of non-significant Results.* Results of hypothesis tests that are statistically non-significant (and thus non-surprising) are unlikely to be published. This situation may have serious course sequences. Consider a null hypothesis that is in fact true and that is tested independently by many outlays at 5% level of significance.

Each of these investigators has only 5 chances out of 100 of (incorrectly) finding statistical significance (and falsely rejecting H_0^1). Yet the chance that at least one among the many will reach such a result is much higher. If there are 10 such investigators, the probability that at least one will falsely reject H_0 equals $1 - (.95)^{10} = .401$ and that persons's (wrong) results will be published.

(2) *Serious violations of assumptions.* Many, hypothesis tests are performed even though the assumptions that would validate the procedure used are not met. There might be a difference, for example, between the statistical (target) population about which inferences are to be made and the population that is actually being sampled (This difference is likely because a perfect frame is usually unavailable for any non-trial population. In addition, an intended simple random sample can in fact turn out to be a non-probability sample (this result is likely because high non-response rates are common) or, perhaps, the populations being samples may not be normally distributed. or they may not have equal variances (requirements for using the t-statistic in two-sample tests). In testing situations such as these, the testing procedure will grind out nonsense.

(3) *They do not tell us "why" the difference exists.* Though tests can indicate that a difference has statistical significance, they do not tell us why the difference exists. However, they do suggest the need for further investigation in order to reach definite answers.

(4) *Conclusions are to be given in terms of operabilities and not certainties.* When a test shows that a difference was statistically significant it suggests that the observed difference is probably not due to chance. Thus statements are not made with certainty but with a knowledge of *probability. "Unusual"* events do happen once in a while.

(5) *They should not be used mechanically.* Tests of significance are simply the raw materials from which to make decisions, not decisions in themselves. There may be situations where real differences exist but do not produce evidence that they are statistically signficant or the other way round. In each case it is absolutely necessary to exercise great care before taking a decision.

(6) If we have confidence in a hypothesis, it must have support beyond the statistical evidence. It must have a rational basis. This phrase suggests two suggests two conditions : first, the hypothesis must be 'reasonable' in the sense of concordance with a prior expectation. Secondtly, the hypothesis must fit logically into the relevant body of established knowledge.

The above points clearly show that in problems of statistical significance as in other statistical problems, technique must be combined with good judgment and knowledge of the subject-matter.

LIST OF FORMULAES

Tests of Significance of Attributes :

S.E. of number of success $= \sqrt{npq}$

S.E. of proportion of successes $= \sqrt{\frac{pq}{n}}$

S.E. of difference of proportions $= \sqrt{pq\left(\frac{1}{n_1}+\frac{1}{n_2}\right)}$

where $p = \frac{x_1 + x_2}{n_1 + n_2}$

Test of Significance of Large Samples :

$$S.E._{\overline{X}} = \frac{\sigma}{\sqrt{n}}$$

$$S.E._{(\overline{X}_1-\overline{X}_2)} = \sqrt{\frac{\sigma_1^2}{n_1}+\frac{\sigma_2^2}{n_2}}$$

$$S.E._{(\sigma_1-\sigma_2)} = \sqrt{\frac{\sigma_1^2}{n_1}+\frac{\sigma_2^2}{2n_2}}$$

$$S.E._r = \frac{1-r^2}{\sqrt{n}}$$

Tests of Significance of Small Samples :

(1) To test the significance of the mean of a random sample

$$t = \frac{\left(\overline{X}-\mu\right)}{S}\sqrt{n}$$

where $v = n_1 + n_2 = 2$

(2) To test the significance of the difference of the means of two samples

$$t = \frac{\left(\overline{X}_1-\overline{X}_2\right)}{S} \times \sqrt{\frac{n_1 n_2}{n_1+n_2}}$$

where $v = n - 1$

(3) Two test the difference of the means of two samples which are not independent.

$$t = \frac{\overline{d}\sqrt{n}}{S}$$

where $v = n - 1$

(4) To test the significance of an observed correlation coefficient

$$t = \frac{r}{\sqrt{1+r^2}} \times \sqrt{n-2}$$

where $v = n - 2$

The Variance Ratio Test–F-Test :

$$F = \frac{S_1^2}{S_2^2}$$

where $S_1^2 = \frac{\Sigma\left(X_1 - \overline{X}_1\right)^2}{n_1 - 1}$ and

$$S_2^2 = \frac{\Sigma\left(X_1 - \overline{X}_1\right)^2}{n_2 - 1}.$$

EXERCISES

1 In a sample of 50, there are found to be 27 males and 23 females. Ascertain if the observed proportions are inconsistent with the hypothesis that the male and female should be in equal proportion.

2. What do you mean by test of significance of a mean or difference between two means ? Define standard error. Explain and mention formula to obtain standard error of mean and standard error of standard deviation both in ungrouped and grouped data.

3. State the meaning and utility of standard error. Describe briefly the procedure for carrying out a hypothesis testing.

4. What is sampling distribution of the Mean? State its important properties.

5. Distinguish between Estimation and Estimator. State the desirable properties of a good estimator.

6. Describe the various steps involved in testing of a hypothesis. What is the role of standard error in testing hypothesis?

7. Two different types of drugs A and B were tried on certain patients for increasing weight. Five persons were give drug A and seven persons were given drug B. The increase in weight in pound is given below :

Drug A :	8	12	13	9	3		
Drug B :	10	8	12	15	6	8	11

Do the two drugs differ significantly with regard to their effect in increasing weight? (table value of t for v = 10 at 5% level is 2.23).

8. A sample of 400 persons was taken from a large population. The mean weight and standard deviation of these persons was found to be 68 kg. and 6 kg. Can it reasonably be said that in the population the mean weight would be 67 kgs?

9. A large corporation employs both men and women to do the same kind of work. It is hypothesised that the average hourly output of men is less than that of women. The following information was gathered by the O and M teams of the corporation :

	Men	*Women*
Variance in units	$\sigma 1^2 = 70$	$\sigma 2^2 = 74$
Sample size	$N_1 = 36$	$N_2 = 36$
Sample mean in units	$\overline{X}_1 = 150$	$\overline{X}_2 = 153$

Is the average hourly output of men significantly lower than that of women at 5% level of significance?

10. In one section of a large city, only 200 out of 250 residents own the dwellings in which they live. In other section, 120 out of 200 residents own their homes. Using a 0.05 level of significance, can we conclude that there is no difference between these two section of the city in terms of house ownership?

11. An examination was given to two classes, each consisting of 49 students. In the first class, the mean marks were 74 with a standard deviation of 8, in the second class, the mean marks were 78 with standard deviation 9. Is it reasonable that two classes are having been obtained from two normal populations with equal means? Test your hypothesis at 5% and 10% level of significance.

12. The breaking strengths of cables produced by a manufacturer have been 1800 lbs-and standard deviation 200 lbs. By a new technique in the manufacturing process, it is claimed that the breaking strength can be increased. To test this claim, a sample of 64 cables is tested and it is found that the mean breaking strength is 1850 lbs. Can we support the claim at (i) 5 per cent (ii) 1 per cent levels of significance?

13. A large hotel chain is trying to decide whether to convert more of its rooms to non-smoking rooms. In a random sample of 400 guests last year, 16 had requested non-smoking rooms. This year, 205 guests in a sample of 380, preferred the not-smoking rooms.

World you recommend that the hotel chain converts more rooms to non-smoking? Support your recommendation by testing the appropriate hypothesis at 0.01 level of significance?

14. (a) State the procedure followed in testing a hypothesis.

    (b) Explain the procedure generally followed in testing of a hypotheses. Point out the difference between one tail and two tail tests.

15. (a) Define null·hypothesis, critical region and two sided test used in testing of hypothesis.

    (b) Explain the concept of standard error. How is it useful in testing of hypothesis?

16. Discuss the role of hypothesis in making a research design. What are the characteristics of a good hypothesis?

17. Two independent random samples of size 10 and 15 from two independent normal populations having variance 4 and 8 were found to have means 48 and 46 respectively. Test at 5% significance level whether the population means may be taken to be equal. What assumptions do you make in applying the test? Also find 95% confidence limits for thew difference of the population means.

18. The following data show weekly sales of a manufacturer BEFORE and AFTER recognition of the sales organisation, 10 weeks from Sept, to Dec. in two successive years were selected for comparison:

Week No. :	1	2	3	4	5	6	7	8	9	10
sales (before Reorganisation) (in '000 Rs.) :	15	17	12	18	16	13	15	17	1918	
Sales (After Reorganisation) (in '000 Rs.) :	20	19	18	22	20	19	21	23	24	24

Apply the 't' test to determine whether re-organisation had any effect on the sales.

[t = 11.86]

19. In a sample of 100 respondents it is found that 55 view a T.V. Serial. Find a 99% and 95% confidence interval for the population proportion.

20. As a financial manager, you are considering two different television advertisements for promotion of a new financial service. Two test market areas with identical consumer characteristics are selected. Advertisement A is used in one area and B is used in the other

area. In a random sample of 60 customers who saw advertisement A, 18 used the service. In another sample of 100 customers who saw advertisement B, 22 tried the new service. Would you say that advertisement A is more effective than advertisement B at 5% level of significance?

21. Of the two salesmen, X claims that he has made larger sales than Y. For the accounts examined, which were comparable for the two men, the results were:

	X	*Y*
Number of sales	10	17
Average size	Rs. 6,200	Rs. 5,600
Standard deviation	Rs. 690	Rs. 600

Do these two 'average size of sales' figures differ significantly? Explain your result. You may use the following extract from the table of Student's t-distribution.

22. A random sample of 200 villages from Jaipur district was taken and average population per village was found to be 420 with 50 as standard deviation from the same district, another sample of 200 village as 480 and 60 as standard deviation. Is the difference between the average of two samples significant?

23. A mean buys 200 electric bulbs of each of two well-known makes taken at random from stock for testing purposes. He finds that 'Make A' has a mean life of 2,560 hours with a standard deviation of 90 hours and 'Make B' has mean life of 2,650 hours with a standard deviation of 75 hours. Is there a significant difference in the mean life of these two makes at 5% level of significance?

24. What do you understand by t-test? Explain. A sample of 11 chickens of two months old was taken which was reared on Hind Lever diet. Weights were found to be 14, 16, 14, 16, 18, 16, 18, 16, 17, 14 and 17 ounces. Another sample of 11 chickens of the same group was taken which was kept on another diet. Their weights were fond to be 12, 14, 13, 14, 16, 17, 16, 17, 16, 14 and 16 ounces. Test whether two diets are significantly different? The table value of t for 20 d.f., at 5% level of significance is 2.086.

25. The wages of 10 workers taken at random from a factory are given below :

Wages (Rs.):	578,	572,	570,	568,	572,
	578,	570,	572,	596,	584

Is it possible that the mean wage of all workers of this factory is Rs. 580? (Given for v = 2, t = 2.26

26. A random sample of 225 families in a city indicates that they average 150 minutes of television viewing per day with a standard deviation of 30 minutes. What are the 99 per cent confidence limits for the average viewing time of all families in this city?

27. A sample of 900 members is found to have a mean 3.4. Can it be reasonably regarded as a simple sample from a large population with mean 3.25 and standard deviation 2.61?

28. (a) 10 tests of two groups, viz., Swaraj and Swadeshi. Give the following results.

Sawaraj	:	mean = 64	S.D. = 8	n = 100
Swadeshi	:	Mean = 61	S.D. = 10	n = 80

(i) Is the difference in mean scores significant?

(ii) Is the difference in standard deviations significant?

[(i) d.f./S.E. = 2.18 (ii) Diff./S.E. = 2.06]

29. It is hypothesized that more than 10 per cent of the families in the United States plan to buy new automobiles this year. A sample of 2,000 families interviewed in a nationwide survey, indicated that 160 planned to purchase new cars this year. Do you accept the hypothesis at 5 per cent level of significance?

30. The annual salary of administrators in the Government averages Rs. 20,000 and has a standard deviation of Rs. 1,500. In the same Government, the salary of doctors averages Rs. 22,500 and has a standard deviation of Rs. 2,250. These data relate to samples of sizes of 25 for each group chosen at random. Test at 5% level of significance whether there is significant difference in the mean salaries of two groups.

31. Daily sales figures of 40 shopkeepers were collected and the average sales and standard deviation were fond to be Rs. 528 and Rs. 600 respectively. Is the assertion that daily sale on the average is Rs. 400 only, contradicted at 5% level of significance by the sample?

32. (a) Tests were made at short intervals on spark plugs from two manufacturers. The following table gives the number of hours of service given by plugs from the two sources :

A :	200	210	190	200	190	200	180	200	200	210
B :	190	200	210	190	180	190	200	192		

Do these results indicate a statistically significant difference between the spark plugs os far as mean length of service is concerned?

(b) Intelligence test on two groups of boys and girl gave the following result :

	Mean	*S.D.*	*N*
Girls	75	15	150
Boys	70	20	250

Is there a significant difference in the mean scores obtained by boys and girls?

33. What is a "Point estimator"? Give examples. How will you proceed to construct an interval estimate of the mean of normal population with known variance?

34. (a) Why should there be different test functions for testing the significance of difference in means when samples are (a) small, (b) large?

(b) Explain the term "Standard Error" and show how standard error is used in large sample test of statistical hypothesis.

35. (a) Explain the terms : (i) Type I and Type II errors; (ii) Critical region; and (iii) Level of significance associated with testing hypothesis.

(b) Explain the difference between the 'standard error of mean' and 'standard error of proportions'.

36. (a) Explain the concept of sampling distribution in general and also explain the sampling distribution means in particular. What are its properties?

(b) Explain Type I and II errors associated with testing of hypothesis. How will you proceed to test the difference between the means in two population, based on large samples drawn from the populations?

37. Explain the procedure for testing the hypothesis concerning the difference between two population proportions based on samples taken from each of the two populations.

38. On the basis of their total scores, 200 candidates of a civil service examination are divided into two groups, the upper 30 per cent and the remaining 70 per cent. Consider the first question of this examination. Among the first group, 40 had the correct answer.

On the basis of these results, can one conclude that the first question is not good at discriminating ability of the type being examined here?

39. A company is interested in knowing if there is a difference in the average salary received by foremen in two divisions. Accordingly, samples of 12 foremen in the first division and 10 foremen in the second division are selected at random. Based upon experience foremen's salaries are known to be approximately normally distributed, and the standard deviations are about the same.

	First Division	*Second Division*
Sample size	12	10
Average monthly salary of foremen (Rs.)	1,050	980
Standard deviation of salaries (Rs.)	68	74

40. A survey of 17 agricultural labourers reveals an income of Rs. 40 per week with a standard deviation of Rs. 8. Find to the limits of mean weekly wages in the population at 5% level of significance.

41. (a) Two types of batteries X and Y are tested for their length of life and the following results are obtained :

Battery	*Sample size*	*Mean hours*	*Variance (hours)*
X :	10	1000	100
Y :	12	1020	121

Can you conclude that the two types of batteries are having the same mean life?

(b) Test the significance of the difference between the means of the sample from the following data :

	Size of sample	*Mean*	*S.D.*
Sample A	100	50	4
Sample B	150	51	5

42. (a) What is t-distribution? Explain some of its applications.

(b) What is t-distribution? Explain its uses in testing the hypothesis.

(c) Explain how Student's 't' test is used to test the significance of the difference between the means of two samples?

43. Differentiate the following pairs of concepts :
    (i) Statistics and parameter,
    (ii) Critical region of acceptance, and
    (iii) Null and alternative hypothesis.

44. The management of a company claims that the average weekly income of their employees is Rs. 900. The trade union disputes this claim stressing that it is rather less. An independent survey of 150 randomly selected employees estimated the average to be Rs. 856 and the Standard Deviation to be Rs. 364.26. Would you accept the view of the management or the trade union?

    [Diff./S.E. = 1.48]

45. Explain how the Student's 't' distribution may be used to :
    (i) test the significance of the sample correlation coefficient in a sample drawn from a bivariate normal population.
    (ii) test the significance of the difference between the yield of two varieties in agricultural experiment.
    (iii) explain Fisher's transformation of the correlation coefficient and indicate its use in tests of significance.
    (iv) explain briefly different applications of the t-test.

46. (a) Difference between standard deviation and standard error of estimate.
    (b) Explain the following :
    (i) Null hypothesis.
    (ii) Statistical hypothesis
    (c) Explain briefly the need for sampling in the context of estimation of unknown population parameters and statistical testing.
    (d) Explain the concept of degrees of freedom.
    (e) Write short notes on :
    (i) Type I and Type II errors
    (ii) Steps involved in testing of a hypothesis.

47. (a) What is null hypothesis?
    (b) Distinguish between large sample and small sample.

(c) State the limitations of tests of significance.

(d) Write a note on the sampling distribution of the Sample Mean.

48. Suppose you want to estimate the proportion of families in your town which has two or more children. A random sample of 144 families shows that 48 families have two or more children. Construct a 95% confidence interval.

49. (a) Define student's t-statistic and write its probability density function.

(b) Explain the procedure that is followed i testing of a statistical hypothesis.

(c) What do you mean by estimation? Discuss the desirable properties of an estimator.

50. (a) Explain the terms Sampling distribution and Standard error of a statistics.

(b) Show that mean of the Sampling distribution of the Sample mean is equal to the population mean.

51. A certain stimulus administered to each of 12 patients resulted in the following changes in blood pressure :

5, 2, 8, –1, 3, 0, –2, 15, 0, 4, 6.

Can it be concluded that the stimulus will, in general, be accompanied by an increase in blood pressure?

52. Eleven school boys were given a test in science. They were given a month's special coaching and a second test was held at the end of it. Do the marks give evidence that the students have benefited by the extra coaching?

Marks in first test	:	46	40	38	42	36	40
Marks in Second test	:	48	38	44	36	40	44
Marks in first test	:	30	34	46	32	38	
Marks in second test	:	40	40	46	40	32	

[For v = 10, the table value of t at 5% level of significance = 2.23]

53. The incomes of a random sample of engineers in Factory 1 are Rs. 630, 650, 680, 690, 710 and 720 per month. The incomes of a similar sample in Factory II are Rs. 610, 620, 660, 690, 700,

710, 720 and 730 per month. Discuss the validity of the opinion that Factory 1 pays its engineers much better than Factory II.

54. A sample of 400 items is taken from a population whose standard deviation is 1.5. The mean of the sample is 2.5. Test whether the sample has come from a population with mean 26.8. Also calculate the 98% confidence limits of the population mean.

55. A coin is tossed 10,000 times and the head appeared 5,195 times. Would you consider the coin biased?

56. A correlation coefficient based on a sample of size 51 was computed to be 0.32. Use 't' test to find whether it is significant. The table value of 't' for 49 degrees of freedom at 5% level of significance is 2.01.

57. Sky Packets guarantee that 90 per cent of their deliveries are on time. In a recent week, 81 deliveries were made of which 6 were late. Sky packets, Managing Director says with 95% confidence that there has been a significant improvement in deliveries. Should managing Director's statement be accepted?

58. A random sample of 100 mill workers at Kanpur showed their mean wages to be Rs. 350 per week with a standard deviation of Rs. 28. A sample of 150 mill workers in Bombay showed the mean wage to be Rs. 390 per week with a standard deviation of Rs. 40. On the basis of the data, would you say the mean wages of mill workers in Bombay are higher that those at Kanpur?

59. the mean height obtained from a random sample of size 100 is 64 inches. The standard deviation of the height distribution is known to be 3 inches. Set up 95% limits of the mean height of the population.

60. (a) A die is thrown 9,000 times and a throw of 3 or 4 is reckoned as a success. Suppose that 3,240 throws of a 3 or 4 have been made out. Do the data indicate an unbiased die?

(b) A random sample of size 7 from a normal population gave a mean 977.51 and a standard deviation 4.42. Find a 95% confidence interval for the population mean.

(c) Intelligence test of two groups of boys and girls gave the following results :

Girls	12	84	10
Boys	8	81	12

Examine if the difference of means is significant.

61. A random sample of 400 members is found to have a mean of 4.45 cm. Can it be regarded a sample from large population whose mean is 5 cm. and whose variance is 4?

62. Random samples of 200 bolts manufactured by machine A and 100 bolts manufactured by machine B showed 19 and 5 defective bolts respectively. Is there a significant difference between the performance of the two machines?

63. A departmental store executive reads in a trade journal that a large departmental store has made a survey indicating that over 50% of the persons who enter the store as apparently potential customers,, actually figure and decide to try to see what the percentage of non-buying visitors is for his store. Plans are made calling for discrete surveillance of a random sample consisting of 100 persons, entering and leaving the store. It is found that 41 of these persons made no purchases Test at 5% level of significance, the hypothesis that the percentage of non-buyers is 50.

64. 500 units from a factory are inspected and 12 are found to be defective. 800 units from another factory are inspected and 12 are found to be defective. Can it be concluded at 5% level of significance that production at second factory is better that in first factory?

65. The following table presents data on the value of a harvested crop stored in the open and inside a godown:

Sample	*size*	*Mean*	$\sum(\overline{X}-X)^2$
Outside	40	117	8,865
Inside	100	132	27,315

Assuming that the two samples are random and they have been drawn from normal population with equal variances, examine if the mean value of the harvested crop is affected by weather conditions.

66. (a) Below are given the gains in weights (lb.) of lions on two diets X and Y : Gain in weight (lb.)

Diet X :	25	32	30	32	24	14	32			
Diet Y :	24	34	22	30	42	31	40	30	32	35

Test, at 5% level, whether the two diets differ significantly with regard to increase in weight.

67. A sample of 10 measurements of the diameter of a sphere gave a mean X = 4.38 inches and a standard deviation 0.06 inches. Find (a) 95%, and (b) 99% confidence limits for the actual diameter.
68. Intelligence test given to two groups of boys and girls gave the following information:

	Mean Score	*S.D.*	*Number*
Girls	75	10	50
Boys	70	12	100

Is the difference in the mean scores of boys and girls statistically significant?
69. 500 articles from a factory are examined and found to be 2 per cent defective. 800 similar articles from a second factory are found to have 1.5 per cent defectives. Can it reasonably be concluded that the products of the first factory are inferior to those of the second?

	Sample size	*Sample mean*	*Sample variance*
Yarn A	4	50	42
Yarn B	9	42	56

The strengths are expressed in pounds. Is the difference in mean strengths significant of real difference in the mean strengths of the sources from which the samples are drawn?
70. Distinguish between the one tailed tests and two tailed tests, clearly bringing out the difference between the alternative hypothesis under the two types of tests.
71. Rate of oxygen consumption in 15 groups of fishes of micrographs was studied as 62. 64, 67, 68, 61, 72, 71, 69, 67, 64, 62, 60, 62, 64 and 67 kg/hour/I OOc.c. Obtain standard error of mean and standard error of standard deviation both in ungrouped and grouped series.
72. What is the major purpose of hypothesis testing? Explain the various steps involved in hypothesis. What is sampling distribution? Discuss its utility in significance tests.
73. How does the standard error of the mean measure sampling error? Is the amount of sampling error in the sample mean affected by the amount of variability in the universe? Explain. Why is it important to consider sampling error?

74. What is 'test of significance'? Explain the concept of standard and discuss its role in large sample theory.

75. (a) Why is testing of hypothesis at all necessary?

(b) Define type I and type II error.

(c) Discuss the following in proper details :

(i) Maximum likelihood method.

(ii) Estimation of population variance.

(iii) Point vs, interval estimates.

(d) (i) Explain the Central Limit Theorem and its usefulness.

(ii) Explain some important properties of estimators.

76. Eleven sales executive trainees are assigned selling jobs right after their recruitment. After a fortnight they are withdrawn from their field duties and given a month's training for executive sales. Sales executed by them in thousands of rupees before and after the training in the same period, are listed below.

Sales ('000Rs.) (Before training) :	23	20	19	21	18	20	18	17	23	16	19
Sales ('000Rs.) (Before training) :	24	19	21	18	20	22	20	20	23	20	27

Do these data indicate that the training has contributed to their performance?

77. In a sample of 600 students of a certain college, 400 are found to use dot pens. Test whether the two colleges are significantly different with respect to the habit of using dot pens.

78. Explain the concept standard error. How is it useful testing of hypothesis?

79. Two types of batteries A and B are tested for their length of life and the following results are obtained :

Battery	*Sample size*	*Mean (hrs.)*	*Variance (hrs.)*
A	10	500	110
B	12	560	121

Is there a significant difference in the two sample means?

80. Name the test of significance which is most appropriate in following situations :

(i) For judging the significance of differences of more than two samples means at one and the same time.

(ii) For comparing the sample variance to a specified variance of the population.

(iii) For comparing the mean of a random sample to some hypothesised mean for the population when 15 and population variance is known.

(iv) For comparing the variances of two independent random samples.

61. What do you understand by-test? Indicate some practical application of t-test.

82. A marketing research analyst collects data for a random sample of 100 customers out of the 400 who purchased a particular coupon special'. The 100 people spent an average of Rs. 2457 in the store with a standard deviation of Rs. 660. Using 95% confidence interval estimate (i) the mean purchase amount for all 400 customers and (ii) the total amount (in Rs.) of purchases by the 400 customers.

83. (a) What is it that which the standard error of estimate measures?

(b) What are Type I and Type II errors in tests of hypothesis. How is a test of hypothesis constructed?

63. (a) Discuss some important application of t-test in making business decisions.

(b) Explain the role of standard error in large sample tests.

(c) What are the main characteristics of normal probability curve? Distinguish between standard normal distribution and t-distribution.

84. (a) A random sample of 16 values from a normal population showed a mean of 48.5 and sum of squares of deviations from the mean equal to 135. Can it be assumed that the mean of the population is 43.5?

(b) A random sample of 12 pairs of observations from a normal population gives a coefficient of correlation of 0.54. Is this value significant of correlation in the population?
[t=1.597]

85. (a) The mean life time of 400 bulbs produced by a company is found to be 2.570 hours with a standard deviation of 120

hours. Test the hypothesis that the alternative hypothesis is greater than 2,6000 hours at the 5 per cent level of significance.

(b) In an examination in Psychology. 12 students in one class had a mean grade of 78 with a standard deviation of 6, while 15 students in another class had a mean grade of 74 with a standard deviation of 8. Is there a significant difference between the means of the two groups?

86. (a) In a random sample of 600 males in Jaipur, 400 were found to be smokers while in another random sample of 900 females in Delhi, 450 were found to be smokers. Discuss the question whether the data reveal a significant difference in Jaipur to Delhi so far as the proportion of smokers is concerned.

(b) I.Q. Test on two groups of boys and girls gave the following results:

Girls	$\overline{X}$=78	S.D.=10	N=50
Boys	$\overline{X}$=73	S.D.=15	N=100

Is there a significant difference in the mean scores of boys and girls?

87. (a) A company produces two makes of bulbs, A and B, 200 bulbs of each make were tested and it was fond that make A has mean life of 2,560 hours and S.D. 90 hours: whereas make B has 2,650 hours of mean life with S.D. 75 hours. Is there a significant difference between the mean life of two makes?

(b) A stenographer claims that she can take dictation at the rate of 120 words per minute. Can we reject her claim on the basis of 100 trails in which she demonstrate a mean of 116 words with a standard deviation of 15 words. Use 5 per cent level of significance.

88. (a) You are given the following information relating to purchase of bulbs from two manufacturers A and B:

Manufacture	*No. of Bulbs bought*	*Mean live*	*S.D.*
A	100	2,950 hours	100 hrs.
B	100	2,970 hours	90 hrs.

Is there a significant difference in the mean life of two makes of bulbs?

(b) A random sample of 500 parts of an engineering product was taken from a large consignment and 65 were found to be defective. Estimate the proportion of the defective parts in the consignment. Also deduce that the percentage of defective parts in the consignment certainly lies between 8.5 and 17.5.

89. A random sample from 200 villages was taken from Meerut district and the average population per village was found to be 250 with S.D. of 50. Another random sample of 200 villages from the same district gave an average population of 580 per village with S.D. of 60. Is the difference between the averages of the two samples statistically significant

90. An automatic machine was designed to pack exactly 5 kilograms of oil. A random sample of 15 tins was examined to test the machine. The average weight was fond to be 4.94 kg. with standard deviation 0.10. Assuming normal distribution for the weights of tine packed, test at 5 per cent level of significance, whether the automatic machine is working properly.

91. Ten specimens of copper wires drawn from two large lots have the following breaking strength (in kg. wt.):

578, 572, 570, 568, 512, 578, 570, 572, 569, 548

Test whether the mean breaking strength of the lot may be taken to be 578 kg. wt.

92. 250 oranges are taken at random from a basket and 25 are found to be bad. Estimate the proportion of bad oranges in the basket and assign limits within which the percentage most probably lies.

93. A machine puts out 16 imperfect articles in a sample of 500. After the machine is over-handed, it puts out 3 imperfect articles in a batch of 100. Has the machine improved?

94. Two groups of students are given an intelligence test (X) and an arithmetic test (Y).

$n_1 = 45$; $r = 0.45$

$n_2 = 39$; $r = 0.08$

Is the difference between the value of r significant?

95. The average hourly wage of a sample of 150 workers in a plant 'A' was Rs. 2.56 with a standard deviation of Rs. 1.08. The

average wage of a sample of 200 workers in plant 'B' was Rs. 2.87 with a standard deviation of Rs. 1.28 Can an applicant safely assume that the hourly wages paid by plant 'B' are higher than those paid by plant 'A'?

96. Tick the correct answer :

(a) Standard error of number of successes is given by :

(i) $\sqrt{\frac{pq}{n}}$ (ii) $\sqrt{npq}$ (iii) npq (iv) $\sqrt{\frac{np}{q}}$ (v) n_2p_2

(b) 95% fiducial limits of population mean are :

(i) $\overline{X} \pm 3 S.E.$ (ii) $\overline{X} \pm 2.51 S.E.$ (iii) $\overline{X} + 2.58 S.E.$

(iv) $\overline{X} \pm 1.96 S.E.$ (v) $\overline{X} \pm 1.95 S.E.$

(c) 99% fiducial limits of population mean are:

(i) $\overline{X} \pm 2.58 S.E.$ (ii) $\overline{X} \pm 1.96 S.E.$ (iii) $\overline{X} \pm 3 S.E.$

(iv) $\overline{X} \pm 2 S.E.$ (v) none of these.

(d) Large sample theory is applicable when:

(i) N > 30 (ii) N < 30 (iii) N = 30 (iv) N is at least 100 (v) N is at least 1,000.

(e) Student's 't' Distribution was discovered by :

(i) Karl Pearson (ii) Laplace (iii) Fisher (iv) Gosset (v) None of these.

(f) The difference of two means in case of small samples is tested by the formula :

(i) $t = \frac{\overline{X}_1 - \overline{X}_2}{S}$ (ii) $t = \frac{\overline{X}_1 - \overline{X}_2}{S}\sqrt{\frac{n_1 + n_2}{n_1 n_2}}$

(iii) $t = \frac{\overline{X}_1 - \overline{X}_2}{S}\sqrt{\frac{n_1 n_2}{n_1 + n_2}}$ (iv) $t = \sqrt{\frac{n_1 n_2}{n_1 + n_2}}$

(v) $t = \frac{\overline{X}_1 - \overline{X}_2}{S}\sqrt{\frac{n_1 - n_2}{n_1 n_2}}$

(g) While testing significance of the difference of two sample means in case of small samples, the degree of freedom is calculated by:

(i) $v = n_1+n_2$ (ii) $v = n_1—1$ (iii) $v = n_1+n_2—1$ (iv) $v = n_1—n_2+2$ (v) $v = n_1+n_2—2$

[a (ii), b (v), c (i), d (i), e (iv), f (iii), g (v)]

97. Fill in the blanks :

(i) The null hypothesis asserts that there is no true difference in the .... and the .... in the particular matter under consideration.

(ii) Type I error is committed when the hypothesis is true but our test .... it.

(iii) Type II errors are made when we accept a null hypothesis which is ....

(iv) In .... tail test rejection region is located in one tail.

(v) The standard deviation of sampling distribution is called ....

(vi) The distribution formed of all possible values of a statistics is called the ....

(vii) The mean of sampling distribution of means is equal to the ....

(viii) Standard error provides an idea about the .... of sample.

(ix) An estimator is said to be .... if it covers as much information as possible about the parameter which is contained in the sample.

(x) An .... estimate of a population parameter provides two values, between which it is estimated that the parameter lies.

[(i) Sample, population (ii) rejects (iii) not true (iv) one (v) standard error (vi) sampling distribution (vii) population mean (viii) unreliability (ix) sufficient (x) interval]

98. Which of the following statements are True or False:

(i) H_a represents the null hypothesis and H_0 alternative hypothesis.

(ii) When the hypothesis is true and our test accepts it. This is called Type I error.

(iii) The reciprocal of standard error is a measure of reliability of precision of the sample.

(iv) An estimator is said to be biased. If its expected value is identical with the population parameter being estimated.

(v) A point estimate is a single number which is used as an estimate of the unknown population parameter.

(vi) An interval estimate should generally be preferred to point estimate.

(vii) The term standard error of the mean is used to refer to the standard deviation of the distribution of sample means.

(viii) The sampling distribution has a standard deviation equal to the population standard deviation divided by the sample size.

(ix) The sample mean $\overline{X}$ is the best estimator of the population mean μ.

(x) A t-distribution is lower at the mean and higher at the tail than a normal distribution.

[(i) F (ii) F (iii) T (iv) F (v) T (vi) T (vii) T (viii) F (ix) T].